HOW WE BECAME SENSORIMOTOR

How We Became Sensorimotor
Movement, Measurement, Sensation

- - - - - - -

MARK PATERSON

 University of Minnesota Press
Minneapolis
London

Chapter 2 was published in a different form as "On Pain as a Distinct Sensation: Mapping Intensities, Affects, and Difference in 'Interior States,'" *Body and Society* 25, no. 3 (2019): 100–35; copyright 2019 SAGE Publications; reprinted with permission. Portions of chapter 3 were published in "Architecture of Sensation: Affect, Motility, and the Oculomotor," *Body and Society* 23, no. 1 (2017): 3–35; copyright 2017 SAGE Publications; reprinted with permission. Portions of chapter 6 were published in "Motricité, Physiology, and Modernity in *Phenomenology of Perception*," in *Understanding Merleau-Ponty, Understanding Modernism*, ed. Ariane Mildenberg, 170–84 (London: Bloomsbury Academic, 2019); reprinted with permission of Bloomsbury Academic.

Published by the University of Minnesota Press
111 Third Avenue South, Suite 290
Minneapolis, MN 55401-2520
http://www.upress.umn.edu

ISBN 978-1-5179-0999-4 (hc)
ISBN 978-1-5179-1000-6 (pb)
A Cataloging-in-Publication record for this book is available from the Library of Congress.

Printed in the United States of America on acid-free paper

The University of Minnesota is an equal-opportunity educator and employer.

UMP BmB 2021

CONTENTS

From Nineteenth-Century Physiology to Twenty-First-Century Neuroprosthesis

Unlike sight or hearing, where a single identifiable organ is associated with that sense, we are less certain of the various kinds of sensation that arise within the bodily interior. As the founder of psychophysics Gustav Fechner put it in 1860: "Objective sensations, such as sensations of light and sound, are those that can be referred to the presence of a source external to the sensory organ," whereas "changes of the common sensations, such as pain, pleasure, hunger, and thirst, can, however, be felt only as conditions of our own bodies" (1966, 15). A decade earlier, the physician and psychologist Ernst Heinrich Weber had initially bundled these sensations together as "common sensibilities" *(Gemeingefühl)* in his *Der Tastsinn und das Gemeingefühl* (1851, translated as *On the Sense of Touch and Common Sensibility,* 1978). Weber's work on the physiology and psychology of touch inspired Fechner's work to rigorously identify, differentiate, and measure sensations such as touch and pain using specially adapted instruments and devices. As the century progressed, an upsurge of interest in experimental physiology and the beginnings of experimental psychology further focused attention to bodily interiority, including the so-called muscle sense, fatigue, balance, and the body's sense of movement and its position in space, bringing together diverse areas of scientific expertise from figures such as Wilhelm Wundt and his idea of muscular sensation, Charles Bell's proposal of a "muscle sense," Ernst Mach's analyses of sensations and movement, and Charles Sherrington's coinage of "proprio-ception" (the sense of bodily position in space).

As this book will show, the scientific work of the identification and measurement of somatic sensations necessitated an inventiveness not only in the building of specialized instruments and apparatus for empirical research, but also in the conceptual framework, inevitably leading to shifts in terminology between time periods and disciplinary areas. Empirical and conceptual work in this area surges between 1833 and 1945, and includes those early sensory experiments in psychology and psychophysics (by Weber and Fechner in Germany), but also physiology (Sherrington in Britain), neurology (Gordon Head and Henry Holmes in Britain, Kurt Goldstein in Germany), and neurosurgery (the mapping of the human motor cortex by Wilder Penfield in Canada); and the tracking of animal and human locomotion through chronophotography (Eadweard Muybridge in the United States, Étienne-Jules Marey in France), the measurement of fatigue in the laboratory and then the factory (Angelo Mosso in Italy, Jules Amar in France), and a return to the very idea of movement, the physiological concept of "motricity," within embodied consciousness (Maurice Merleau-Ponty in France, who explicitly adopts Head and Holmes's idea of "body schema" in 1945). Accordingly, each chapter focuses upon a particular thematic related to bodily sensation, including the "muscle sense," pain, fatigue, balance, proprioception, and the philosophical uptake of the physiological concept of "motricity." Each chapter engages with the corresponding scientific background, including the design and application of specialized measuring instruments. Successively, a picture of how these scientific discoveries informed new approaches to the body emerges.

Proprioceptive Feedback and the Act of Grasping
Fast-forward to the near-present. In his last few months of office, President Obama made a highly publicized tour of the United States to celebrate the country's progress in science, medicine, and technology. On October 13, 2016, Obama visited Pittsburgh as part of this White House Frontiers tour and shook the robotic hand of a wheelchair-using paraplegic named Nathan Copeland. Copeland had lost the use of both arms and both legs after a car accident left him with spinal cord damage as a teenager in 2004. A brain implant developed through DARPA's (Defense Advanced Research Projects Agency) Revolutionizing Prosthetics program, along with National Science Foundation funding, allows Copeland not only to control the movement of his robotic prosthesis but also to regain a sense of touch, to feel objects and people interacting

with his artificial hand. This was all part of the part of the BRAIN (Brain Research through Advancing Innovative Neurotechnologies) initiative, previously announced by Obama with great fanfare and a $100 million budget in 2013. Obama, clearly enjoying himself and aware of the TV news cameras, turned the robotic handshake into a fist bump, and then said impulsively, "Let's blow it up!" Both mimicked exploding fingers, one with his human hand and the other with a robotic one. Obama turned to the gathered crowd and asked rhetorically: "Does everyone fully grasp what's going on here?" (Alba 2016).

The act of *grasping* is indeed the key. Obama explained to the audience that Copeland was not only directing his robotic hand through conscious intention, but also receiving feedback on his own hand's position in space and its resistance to objects. Previous configurations of robotic prostheses had allowed object manipulation, but without this feedback, often experienced as a component of touch, movements are slow and imprecise. To test whether Copeland actually registered the tactile feedback correctly, researchers blindfolded him and then touched each of the fingers of the robotic right hand (see Figure 1). Each time, Copeland correctly identified the location. As the *Washington Post* reported at the time, Copeland said: "I can feel just about every finger. Sometimes it feels electrical, and sometimes it's pressure, but for the most part, I can tell most of the fingers with definite precision. It feels like my fingers are getting touched or pushed" (Nutt 2016). These words show the difference between the touch involved in grasping and the more familiar ideas of cutaneous touch, or sensations on the skin. Scientists call the self-perception of bodily position in space "proprioception," coined by the physiologist Charles Sherrington in 1906, and it is a necessary but uncelebrated component of our everyday somatic experience, as I discuss in chapter 1. In extremely rare cases of neurological disorders, proprioception is lost completely, and Oliver Sacks's (1986) case study of Christina, "The Disembodied Lady," and Jonathan Cole's (1995) account of "I.W." (Ian Waterman) both show the extraordinary consequences of having to learn how to move and walk again, maintaining control of limbs only through constant visual scanning and with great mental effort and concentration.

The production of proprioceptive feedback through neuroprosthesis therefore offers a startling innovation, one with rippling impacts upon those with full or partial paralysis and for amputees. Although

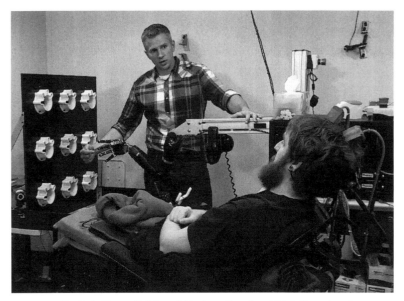

FIGURE 1. *Dr. Rob Gaunt* [left] *and Nathan Copeland* [right] *testing the neuroprosthetic arm and hand. Photograph courtesy of UPMC/Pitt Health Sciences.*

the experiment came together in Pittsburgh, a number of innovations from geographically disparate research institutions had to be developed first. The robotic arm was developed from the Applied Physics Lab at Johns Hopkins in Baltimore, for example, and the microelectrode control system by Blackrock Microsystems in Salt Lake City. DARPA had funded competing neuroprosthetic systems, including a team at Case Western Reserve University under the same Revolutionizing Prosthetics initiative. Furthermore, a related but separate DARPA initiative, HAPTIX (Hand Proprioception and Touch Interfaces), also part of BRAIN, aims "to create fully implantable, modular and reconfigurable neural-interface microsystems that communicate wirelessly with external modules, such as a prosthesis interface link, to deliver naturalistic sensations to amputees" (DARPA.mil). Ethics aside, this surge in military innovation and funding in rehabilitation engineering over the past decade is getting to grips with what has been missing in previous prostheses, the naturalistic experience of proprioceptive as well as tactile feedback. Of course, it also underlines how such feedback is central to the taken-for-granted acts of moving, reaching out, and grasping for every body.

Almost exactly a year before Obama's visit, in October 2015, the principal investigator had invited me to his Rehabilitation and Neural Engineering laboratory to talk about his team's work on this. Dr. Rob Gaunt explained his long-term interest in researching complex sensorimotor function, and that for the DARPA project his team are using a deep brain stimulation (DBS) technique called intracortical microstimulation (ICMS) to convey sensory feedback for fine motor control while avoiding long-term tissue damage. This means that microelectrodes smaller than a grain of sand are implanted as an array into the primary somatosensory cortex (S1), toward the top of the crenelated brain surface. A visible ridge previously known by Sherrington and Penfield as the Rolandic Fissure, now termed the central sulcus, divides the primary somatosensory cortex (sometimes called just the sensory cortex) from the primary motor cortex. Upon receiving signals from the robotic arm, the implanted microelectrodes "artificially evoke tactile percepts," as a paper from the team (Flesher et al. 2016) explains. Intracortical stimulation is often used in experiments with animal subjects, but the results are difficult to verify. Human subjects like Nathan are awake throughout such experiments and able to verbally report the effects of such artificial percepts: "We tracked, over a 6-month period after implantation, the quality of the evoked artificial sensations, the projected locations of these sensations, and the participant's sensitivity to intracortical microstimulation," they summarize (2016, 1). Through rigorous, repetitive trials of moving and interacting with objects in the laboratory, cortical movement representations of the arm, hand, and fingers can be derived. Effectively this is a process of sensory and motoric mapping, methodically tracing reported sensations after stimulation of individual cells by the implanted microelectrode array to specific regions of the hand and fingers. After the mapping and acclimatization process, the subject "feels" pressure in the fingers of their paralyzed right hand when the fingers of the prosthesis are pressed, effectively bypassing their damaged spinal cord. For a wheelchair-using quadriplegic subject who had reluctantly become used to his prosthetic legs, arms, and hands, the ability for Nathan Copeland to regain feeling, tactile feedback, and motor control once again must have been, well, sensational.

The story of Nathan Copeland's celebrity fist bump is built upon the innovations of well-funded teams of neurophysiologists and rehabilitation engineers. But the project of mapping the somatosensory cortex, and the idea of complex sensorimotor function within current research, shows up the legacy of prior physiological investigations at first in the laboratory, and then increasingly out in the field, of the kinds of familiar-feeling sensations that arise through bodily movements, yet which are rarely consciously recognized as such. The purpose of this book is to tell the longer historical narrative behind what evidently continues to be investigated at the forefront of neuroscience: what has been called in the nineteenth century the "sensory-motor" and more recently the "sensorimotor" organization of the body and its perception. In discussion with Gaunt in his lab about the mapping of the somatosensory cortex, there was a readily acknowledged assumption of something like an image, plan, or schema of the experimental subject's sensorimotor body. A comparable assumption was present for a different scientist at a different lab in a different part of the campus a few years later. In November 2019 I arranged a visit to Dr. Gelsy Torres-Oviedo's Sensorimotor Learning Laboratory in the Department of Bioengineering, where she and her team have been working with a professional motion capture ("mo-cap") rig, similar to those used by movie studios, with light-colored balls attached to parts of the experimental subject's dark bodysuit and a series of cameras around a large room. In the center of the large studio-like laboratory is a specially adapted treadmill with two belts, each of which can be independently controlled for speed. The idea is that, if the speed of one belt is altered to be slightly out of sync with the other, this "changes the coordination between the legs, resulting in storage of a new walking pattern," as one of their papers explains (Torres-Oviedo et al. 2011, 65). For the experimental walking subject there is a "locomotor adaptation"—that is, a simultaneous deadaptation from a previously "stored" pattern along with an adaptation to the new altered pattern. When this adaptation process is rendered visible through fMRI imaging, several brain areas light up for conscious adjustments to patterns of leg movement including, as expected, motor regions of the cerebrum, and parts of the cerebellum associated with spatial and temporal control. But what of this idea of pattern "storage" or retention? Habitual movements like walking also involve circuits in the brain stem, and this relationship between conscious movement planning and adaptation, and stored locomotor patterns in the brain stem and spinal cord, offers exciting possibilities to update the cartography of the sensorimotor body.

Torres-Oviedo's work is, like Gaunt's, making eye-opening experimental advances that help us consider the conscious and nonconscious brain and its relationship to the body through movement, then. Both Gaunt's and Torres-Oviedo's work also have clear practical applications for rehabilitation after injury, and for those with acquired or congenital motor impairments. As will be made clear throughout the rest of the book, a synopsis of which is provided below, projects such as theirs can be regarded as a continuation of a long-running concern in the history of physiology and then early neuroscience with the observation, conceptual modelling, neurological mapping, and even mechanical simulation of bodily movement. More particularly, the period between the last half of the nineteenth century and the first half of the twentieth legitimated this area of study scientifically and offered a profusion of innovative instruments and apparatus for the measurement and mapping of indistinct somatic sensations felt within the body, for the graphic inscription of physiological processes such as blood flow and heart rate within the bodily interior, and for the graphic, photographic, and then chronophotographic capture of movements of the bodily exterior. As historians such as Wolfgang Schivelbusch (1986) and Anson Rabinbach (1992), along with the sociologist Zygmunt Bauman (2000), have argued, this same period of modernity is in any case inextricably associated with movement, speed, and acceleration. Concerns and examples that arise within various chapters, such as galloping horses, the high-speed rotation of photographic discs, the factory production line and the effects of industrial fatigue on the body, and the effects of high-speed movement and spinning on the sense of balance, could each be considered as almost paradigmatic instances within this modernist framework of technological acceleration and the fascination with speed. Furthermore, I lean heavily on the history of science throughout. But, as will become steadily clearer the further one proceeds through this book, something else is going on around the edges of the epochal focus. For the introduction of recent examples of scientific inquiry that deal with movement, sensation, and measurement, as with the work of Gaunt and Torres-Oviedo here, is a pattern that recurs throughout each chapter to varying degrees.

The scientific concerns with hitherto underexamined phenomena of the bodily interior, such as pain, fatigue, balance, and the search for a "muscle sense," all crystallize around the same time period, yet the legacy of these bold empirical investigations and the imaginative spirit of the theoretical frameworks that emerge far outlast that era. A few brief examples will suffice. First, there is a direct line from Torres-Oviedo's apparatus for motion capture back to the Paris-based physiologist Étienne-Jules Marey, whose extensive studies on animal and human locomotion culminated in the foundation of a physiological station "in the wild," as it were, outside the walls of his laboratory at the Collège de France, as explored in chapter 4. Marey invents and then develops chronophotography as a method for the effective capture and analysis of bodily motion, leading him to design a black bodysuit for his colleague Demenÿ adorned with regularly spaced silver paper across the limbs. This is clearly the precursor of current "motion capture" technology.

Second, Gaunt's somatosensory mapping through a brain–computer interface (BCI) is a far more precise adaptation at the neural level of an existing technique known as deep brain stimulation (DBS). In chapter 1 the origin of that technique by the pioneering neurosurgeon Wilder Penfield is discussed, where in a parallel to the now-routine Intracortical Microstimulation (ICMS) method, from 1929 onward Penfield at the Montreal Neurological Institute opened up the top of his patients' skull, probed regions of the sensory and motor cortices with an electrically charged instrument, and tried to "map" the cortices by eliciting spoken responses from his non-anesthetized and fully conscious experimental subjects. The aim of these painstaking and difficult neurosurgical investigations for several years was to produce a map of where exactly in the cortices particular sensations around the body were activated. The thumb, the lips, the tongue each had their cortical correlate, for example, and were found after extensive prodding and probing in marathon surgical sessions. Unusually for a busy neurosurgeon, perhaps, Penfield collaborated with a medical artist named Hortense Cantlie to produce an imaginative visualization of the cortical map that became the famous "homunculus," an illustration of the sensory and motor surfaces of the cerebrum with associated body parts and organs draped around it, and which still appears in many introductory psychology textbooks to this day. Although its scientific accuracy has been disputed, and it is in no way as precise as the recent ICMS technique, it nonetheless remains a memorable way to envision the sensory organization of the body. The homunculus as a representation in miniature of the sensory and motor body has its own historical legacy, of course. One of the principal scientists of the era under investigation, Wilhelm Wundt, started his career as a physiologist and became known as the father of experimental psychology in Heidelberg before setting up the first experimental psychology laboratory in Leipzig. In an article of 1862 on the physiology of ocular muscle movements, at one point he makes a more general argument about the purpose of the modelling of physiological behavior:

> The ultimate goal of all investigations in the natural sciences is the artificial, experimental production of processes observed in nature. *The ultimate goal of physiology is the Homunculus.* Even though it is highly improbable, and will always remain a vain wish to put the Homunculus together, scientists have already taken some important steps in that direction. Physiologists will be satisfied without the whole if they only have all the parts in their hands. (Wundt, in Dierig 2003, 128, emphasis added)

Wundt's observation about grasping the parts in order to have a sense of how the physiological mechanisms work is not just an affirmation of the homunculus as a model, but will have echoes for the forms of simulation and modelling of the natural movement in animals (especially birds' wings) and internal organs (including the heart) that Marey, a near-contemporaneous physiologist, would be building with collaborators in Paris a few decades later.

Third, and finally, Gaunt's and Torres-Oviedo's assumption of an internally held map of the motoric body that becomes altered through injury and habilitation has long-standing echoes of a separate, but related, concept to Penfield's (and Wundt's) homunculus from the start of the twentieth century. Discussion with both Gaunt and Torres-Oviedo underlined the continued relevance of an idea originally from Henry Head and Gordon Holmes's 1911 paper in the neurology journal *Brain* of a "schema," or plan, of the body as a whole, derived from their observations of patients with cerebral lesions, and featured in chapter 6. Whereas Penfield's cortical mapping offered a generic and supposedly universal illustration of sensory and motor function based on repeated testing of experimental subjects and their reported experience, Head and Holmes' body schema is compounded from each individual subject's own sensory and motor experience of the world. It was developed by Freud's pupil Paul Schilder as an inherently spatial "body image" and summarized neatly in his *Image and Appearance of the Human Body*: "The image of the human body means the picture of our own body which we form in our mind, that is to say the way in which the body appears to ourselves" (1935 [2013], 11). More technically, it is a mechanism that forms the basis of comparison when our posture is altered or our body is passively moved, as Head and Holmes explain: "At any moment we can become conscious of the position of any part of our bodies, and although such postural recognition is not constantly in the central field of attention, it always forms the measure against which we judge subsequent change" (1911, 186). This postural image, or body schema, is different from a merely visual image, as when one's body moves an immediate reference is made to prior postures, a kind of dynamic and proprioceptive image rather than a visual and static one, which "rises into consciousness as a measured postural change" (187). They continue:

> For this combined standard, against which all subsequent changes of posture are measured before they enter consciousness, we propose the word "schema." By means of perpetual alterations in position we are always building up a postural model of ourselves which constantly changes. Every new posture or movement is recorded on this plastic schema, and the activity of the cortex brings every fresh group of sensations evoked by altered posture into relation with it. Immediate postural recognition follows as soon as the relation is complete. (187)

Meanwhile, along with such postural sensations, having a part of the body touched brings up another "schema" that relates to the surface of our bodies. In other words, the kinds of inquiry we started with, regarding sensations arising within the bodily interior and the multiple forms of "touch," actually bring up a series of "schemata" that interrelate and experientially overlap at times. Crucially, these schemata are not limited to the boundaries of an individual body, as the way that prostheses such as artificial limbs, walking sticks, or other extensions of the body become incorporated into one's body schema over time was of interest to them:

It is to the existence of these "schemata" that we owe the power of projecting our recognition of posture, movement and locality beyond the limits of our own bodies to the end of some instrument held in the hand. Without them we could not probe with a stick, nor use a spoon unless our eyes were fixed upon the plate. Anything which participates in the conscious movement of our bodies is added to the model of ourselves and becomes part of these schemata: a woman's power of localization may extend to the feather in her hat. (188)

Both Head and Holmes's paper and Schilder's book were drawn upon by Maurice Merleau-Ponty in his analysis of *le corps propre* (one's own body) in his *Phénoménologie de la perception* (1945), and are crucial in other recent case studies such as the aforementioned Jonathan Cole's work with "I.W." and his loss of proprioception and touch (1995, 2016), discussed further in chapter 6. The sometimes-problematic implications of utilizing case studies for theorizing "motricity" or the concept of bodily movement as such forms the main concern of that chapter. For now, note that Head and Holmes hypothesize the sensory cortex as the principal organizing area for detecting changes in posture and movement, speculating that it functions through a sweeping attention to periodically changed states of stimuli throughout the body rather than a continual consciousness of body states at all times.

A Note on Methodology

First, this book benefits from an amount of original archival work. A grant in 2016 from the Dietrich School of Arts and Sciences at the University of Pittsburgh allowed me to conduct archival research on the British neurophysiologist Charles Sherrington in London at the Wellcome Library and the UCL Institute of Neurology, and on the Canadian neurosurgeon Wilder Penfield in the Osler Library of the Montreal Neurological Institute. Consequently, original archival materials were consulted, including letters and original pencil-and-ink illustrations. This archival research directly benefits chapter 1 ("The 'Muscle Sense' and the Motor Cortex"), providing added historical depth regarding the "sensory homunculus" of Penfield, and Sherrington's correspondence about the reflex arc and its place in the nervous system.

Second, as demonstrated earlier with the story of Obama and contemporary laboratory work in neuroprosthesis, most chapters will similarly help contextualize the historical focus by referring to a recent scientific study or set of examples. The intention is to demonstrate the continued relevance of a conceptual legacy that originally opened up during this period of physiology and psychology research to the reader.

The rise of physiology as a distinct and reputable science from its subordinate position to anatomy in the second half of the nineteenth century, as explained in chapter 4, is a significant piece of the puzzle. What has been delightful in reading the original papers and monographs of certain physiologists such as Marey and Sherrington, psychophysicists such as Fechner, and neurologists such as Head, Holmes, and Goldstein, is the willingness to hypothesize, to open up a conceptual framework or system based on original empirical observation of life processes. Therefore, while the lasting legacy of the scientists involved is acknowledged, this book's trajectory is less about following a particular strand in the history of science, or the detailed work of intellectual biography. Instead, it traces the long conceptual shadow of a transdisciplinary mode of inquiry toward life processes made possible through a renewed experimental physiology, and how certain discoveries in the laboratory, but also in physiological fieldwork "in the wild," as it were, have shifted the way the body and its interiority—its self-perceptions—are conceived.

Third, several of the chapters involve what is known in current academic parlance as "art–science collaborations." Historical examples of such crossover between the visual arts, especially, and physiological investigations of sensation and movement occur regularly throughout the time period under examination. For example, the 1870s brought attention to the capture of naturalistic movement in painting in the wake of the motion studies of Muybridge and the chronophotography of Marey, as chapter 4 details. So-called "geometric" or "partial" chronophotography rewrote the rules of painted depictions of human and animal movement, most famously in the case of Marcel Duchamp's "Nude Descending a Staircase No. 2" (1912), but the chapter discusses how the effect on earlier painting, including J. M. W. Turner's "Rain, Steam and Speed—The Great Western Railway" (1844) was equally profound. Already mentioned is the series of collaborations between the neurosurgeon Penfield and the medical artist Cantlie from the 1930s until

1957—in fact, with various versions of the homunculus modified for different publications. Chapter 6 on "motricity" brings in trends in modernist painting for the representation of bodies in motion, and a highly apt movement known as "physiological aesthetics" that attempted to offer a truly scientific understanding of artistic effects well before fMRI scans became the norm. The third chapter, on "The Oculomotor," offers sustained examples of built environments through the lens of art historical scholarship, using imaginative walk-throughs of archaeological sites, to evoke a heightened physicality of the movements of the eye and the feet for a visitor navigating that space. In other words, imaginative points of entry are offered to the reader at various points to help translate historically significant scientific ideas into visual representations, either through the efforts and collaborations of the scientists themselves or through my own descriptive methods. At times this will encourage the reader to imaginatively place or project their body, their sensations, even their potential movements into the "scene," perhaps.

Fourth, the role of the burgeoning transdisciplinary field of "sensory studies," and the decades-long cultural–historical work on the senses, should be acknowledged. There is a rich and growing multidisciplinary tradition of scholarship on the senses that spans the humanities and social sciences, and published work and conference panels have acknowledged this "sensory turn" increasingly over the past two decades. Cultural–historical work on the senses has been pioneered by the anthropologist David Howes, editor and author of a number of books on the senses since the 1990s, along with cultural historian Constance Classen, two of whose books embrace touch in particular (2005, 2012). For the most part, this sensory scholarship has concentrated upon the five senses, an arbitrary division that Aristotle settled upon despite the fact that the experience of "touch" is manifold, its physiology diverse and distributed, and that it is inseparable historically from other internally felt sensations (Classen 1993; Heller-Roazen 2007; Paterson 2007, 2016). In the 1940s, biologists discovered other sensory capabilities in animals such as magnetoreception, electroreception, echolocation, and vibrational sensing, but awareness of somatic sensations outside of the usual five sense model in humans has been present in some form even within Aristotle's writing, conceptualized loosely as "inner touch" or "common sensation" before becoming resolved into more scientifically verifiable and distinct perceptual modalities in the nineteenth century, notably "muscle sense," "kinesthesia," and "proprioception." Writing specifically about the scientization of the interoceptive senses in *The Sixth Sense Reader,* Howes makes the comparable claim that "the anatomical basis of the doctrine of the inner senses was discredited by advances in physiology" (2009, 20). Methodologically, then, each chapter echoes the larger trajectory of the book by charting the scientific innovations and physiological experimentation that produce a more distinct and concrete set of modalities. My point in bringing up existing sensory scholarship at this stage is to ward off any assumption that the scientific validation processes described in this book are accepted uncritically. Louise Vinge puts in succinct terms what many cultural historians have long been arguing: "The number and order of the senses are fixed by custom and tradition, not by nature" (1975, 107). In his work, Howes argues likewise that the role that culture plays in shaping sensory experience is independent from any neurobiological explanation of the function of sensory organs. This is the case whether the "culture" in question is that of the Cashinahua indigenous peoples of Peru and Brazil, whose model of the senses is conceived more in terms of "sensorimotor complexes" according to Howes (2009, 25) than as sense modalities as such. Or whether, as I argue in this book, the "culture" is medico-scientific with the emerging respectability of physiology as a discipline in itself, one that takes a turn to the bodily interior in the nineteenth century with Claude Bernard's *milieu intérieur* (1865 [1984]) and the so-called interoceptive sensations of Charles Sherrington (1907). In other words, while there is some continuity with existing cultural–historical work on the five external senses (sight, hearing, smell, taste, and touch) so capably charted by Classen, Jütte (2005), and others, this book represents a departure from that history by shifting attention to the bodily interior, and by tracing those shifts in scientific areas of concern and forms of graphic inscription that brought formerly indistinct bodily sensations into focus and then made them measurable.

Finally, given that previous points already indicate a somewhat trans-disciplinary approach, it is important to get to grips with the methodological trajectory of the book as a whole. Concentrated in, but not exclusively originating from, a remarkably productive historical period of inquiry into sensation and movement within the life sciences, a series of dialogues and debates between German, French, and British psychophysicists, physiologists, psychologists, and neurologists bloomed into more abstract territory, leading to uptake in philosophy and elsewhere. The particular time period is chosen since it is generative of a set of concepts and approaches that emerge and become formalized within the scientific literature, but this is not strictly a work of history, nor even of the history of science. In offering sometimes detailed historical background to concepts such as "proprioception," the "sensory-motor," "pain," "fatigue," and so on, the purpose is to construct a cumulative narrative regarding the identification of sensations arising from the bodily interior and their subsequent means of measurement and mapping. This is not about the production of "knowledge" as such, nor even of the production of particular forms of knowledge about sensation. Instead, my argument and methodology is more akin to a Foucauldian archaeology of physiological *perception*. Let me briefly outline the distinction. In *L'Archéologie du savoir* (1969), Michel Foucault outlines how a whole chaos of signs enters into an "episteme," a universal regime of systems of representation that historically govern and structure the bounds of what counts as knowledge, especially in the human sciences. At various points in this book there are glimpses of what an episteme regarding measurement, the production of "normal" human subjects in terms of distributions and thresholds of sensation in the laboratory and elsewhere, and consequently of deviance, might consist in, especially in the second chapter, on pain, and the fifth chapter, on fatigue. However, a Foucauldian archaeology of knowledge is not interested in a fixed explanation for how certain approaches to knowledge, or what he calls "discursive formations," come about, as they are continually morphing and being reshaped. Foucault argues that an episteme acknowledges this capacity for reshaping, since "it is a constantly moving set of articulations, shifts and coincidences that are established, only to give rise to others" (2002, 211). Unlike a Foucauldian archaeology of knowledge, however, the concern of this book is not to determine the orderly or "unconscious" structures underlying the production of scientific knowledge within a particular time and place. The framework through which the different sections of each chapter emerge, including the scientific studies, case reports, biographical fragments, and archival discoveries, involves a cumulative recognition and then validation of previously vaguely conceived or barely conscious sensations, and the scientific means through which they are measured, recorded, and graphically represented. The framework is therefore ultimately about a shift in institutional, social, and then subjective perception of previously invisible and unrecognized bodily sensations. We might turn instead to Foucault's earlier *The Birth of the Clinic: An Archaeology of Medical Perception* (1963), with its focus on the invention of the anatomo-clinical method and the invention of technologies, expertise, and apparatus to render visible the interior of the body by an accumulation of institutional expertise in the setting of the clinic. It would be a mistake to identify this "medical gaze" merely as remote, detached, even if in the clinic it becomes invested with the power of authority. "The clinical gaze is not that of an intellectual eye that is able to perceive the unalterable purity of essences beneath phenomena," as Foucault explains. "It is a gaze of the concrete sensibility, a gaze that travels from body to body, and whose trajectory is situated in the space of sensible manifestation" (2003, 120). A "sensible truth" does indeed become available, but only to those with the ability to cultivate a "fine sensibility" (121) regarding the interpretation of symptoms as signs. A diagnostic calculus can be performed based not only on visual observation, but also on forms of listening and palpation, for example, in the clinic.

This book is therefore a project of archaeology insofar as it identifies a form of medico-scientific discourse and praxis that identifies, measures, and tracks hitherto underexamined sensations within the body by means of increasingly sophisticated equipment, at first in the laboratory and then in the field. Yet the "sensible truths" being validated do not arise as indexical signs, as experienceable through the cultivation of certain sensibilities in the body of a clinician or a dance instructor, for example. Instead, the observational expertise perhaps more straightforwardly results in ever-finer inscriptive practices, output at first by hand and then though machines into the lines, curves, and graphs that are so recognizable as the products of scientific inquiry in modernity. As we shall see, any new sensible truths become transcribed via the so-called graphic method, which started with Karl Ludwig and Hermann von Helmholtz, and was developed more fully by Étienne-Jules Marey. Methodologically, my project also clearly owes a debt to the work of Princeton historian Anson Rabinbach, whose *The Human Motor: Energy, Fatigue, and the Origins of Modernity* (1990) can be characterized as a novel cultural history of science and technology, and which also depicts the formation and impact of new areas of scientific application. In that vein, my book involves a particular conjunction of historical scientific methods and conceptual approaches to life processes that gathers pace in the second half of the nineteenth century onward. The narrative involves the resurgence of physiology as a science and a resolute experimentalism toward sensations by psychophysicists and physiologists, to be sure, but there are also conceptual ripples, shared influences, and currents from other disciplines. Since academic disciplinary labels and identities were not held so rigidly during the era under observation, the relative reputation of disciplines and their subdisciplines shifts and morphs as the epistemes take form and solidify over time. Herman von Helmholtz was a mathematician and physicist, for example, whose work with the myograph inspired generations of physiologists, as will be seen, including Angelo Mosso's work on fatigue and Marey's generic "graphic method." From Charles Sherrington's physiological observations on reflexes in animals published in 1906 and 1907, as another chapter here details, for example, he is able to construct an influential overarching theory applicable across a number of organisms, and against which neurologist Kurt Goldstein in turn explicitly reacts in his theory of *The Organism* in 1934. The findings of physiologists like Amar and Mosso laid the groundwork for the rigorous scientific investigation of the workplace, and therefore of the science of ergonomics, as well as the forms of time studies of F. W. Taylor and motion studies of Frank and Lillian Gilbreth, who elevated the industrial applications of observations of moving bodies. Lastly, the neurology of Goldstein was hugely influential for the phenomenological psychology of the philosopher Maurice Merleau-Ponty. In other words, the numerous transverse movements and points of connection between academic fields and the artistic world during this historic period are highlighted, as each chapter branches out along roughly similar timelines to focus on a unique sensory modality in formation, tracking scientific pieces of the bigger epistemic picture, as it were. But over the longer arc of the book, a narrative unfolds from the early nineteenth century onward concerning the newfound legitimation of scientific forms of attention to sets of sensations and movements that arise within the bodily interior, displacing the indistinctness of prior introspective methods of investigation by means of rigorous measurement with instruments. The proliferation of graphic means of inscription regarding sensation and movement gathered pace first within laboratories, and then outside in the field with more elaborate apparatus into the twentieth century, culminating in Marey's large scientific field station in Paris's Parc des Princes in 1882 and visits by physiologists to factory production lines in the United Kingdom and the United States around World War I. Along the way there are numerous parallels to current fascinations with movement and the bodily interior, including biometrics and the so-called Quantified Self movement, for example (see, e.g., Lupton 2016, Paterson 2018b). This narrative progression will become clearer in the following section, which offers synopses of the arguments from each chapter.

The Schema, or Plan, of the Book

The first chapter, "The 'Muscle Sense' and the Motor Cortex: A Cartography of Bodily Interiority," involves a somewhat epic journey, starting with hitherto unformed, indistinct somatic sensations known for centuries simply as forms of "common sensation" or "inner touch" (e.g., Heller-Roazen 2007). These came to be identified as more distinct, and consequently measurable, toward the end of the nineteenth century, and were translated into a medicalized terminology. On the one hand, it involves simultaneous physiological investigations mostly in Britain and Germany into what became known for a short while as the "muscle sense," and later "proprioception" and "kinesthesia." On the other hand, the distributed nerve endings in the muscles from which such sensations were found to originate could be traced to their brain-based correlate, the "motor cortex." The "muscle sense" was first coined in German as *Muskelsinn* around 1870, with ongoing discussion by figures such as George, Schäfer, and Goldscheider until the term fell out of favor around 1900. Reports of this work were summarized in contemporary English journals, although Bastian, Ferrier, and Lewes were rather critical of the term. Henry Charlton Bastian's rather thorough investigation of the "muscular sense" included his paper "On the Muscular Sense and the Physiology of Thinking" (1869) and an appendix in his book *The Brain as an Organ of Mind* (1880). However, based on his evaluation of physiology literature, Bastian proposed an alternative term, "kinaesthesia," which referred more specifically to a sense of movement. Bastian's near contemporary, Sir Charles Sherrington, remained unconvinced, and proposed instead a sense of one's own body, "proprioception," in his *Integrative Action of the Nervous System* (1906) and essay

"On the Proprio-ceptive System, Especially in Its Reflex Aspect" (1907). Both Bastian and Sherrington were building on earlier discoveries in Germany, where the *Tastsinn,* or sense of touch, was differentiated from a generalized somatic sensation *(Gefühlssinn)* as Kirchner termed it, or a "common sense" *(Gemeinempfindungen)* for Wundt. Second, key experiments involving the stimulation of monkey brains by Ferrier in 1874 revealed a "motor cortex" that dealt with bodily movement and its planning, which encouraged the American neurosurgeon Harvey Cushing to conduct experiments to map the motor parts of the human brain between 1902 and 1912. It was through building upon the work of Ferrier, Sherrington, and a surgical apprenticeship with Cushing in particular, that led Canadian neurosurgeon Wilder Penfield to create maps of the sensory and motor cortices of the brain that were then depicted in the form of a "homunculus," first in the 1930s with Boldrey (Penfield and Boldrey 1937) and later significantly updated with Rasmussen (*The Cerebral Cortex of Man,* 1950). The resulting "homunculus," a set of schematic illustrations that indicate cortical areas supposedly responsible for sensations ("sensory homunculus") and bodily movements ("motor homunculus"), are reproduced in innumerable psychology textbooks to this day. My archival research at the Montreal Neurological Institute (MNI) involved consulting the original photographs of open-head brain surgery and the ink illustrations that were prepared for these publications by the artist Cantlie. Penfield and his scientific collaborators laid the foundations for what is now called the primary somatosensory cortex ("S1"). Unexpectedly, new light has recently been shed on the story of these illustrations. Paula Di Noto et al. (2013) found that Penfield had mapped both female and male patients, but failed to incorporate the female data into his model. Her team revisited Penfield's archives in order to reclaim the data from female subjects, and were able to produce an alternative to the androcentric homunculus they call a "*her*munculus." The story of this astonishing rediscovery of data, and the challenge to Penfield's previous cartography of the sensory and motor cortex, rounds out this chapter.

Chapter 2, "Pain as a Distinct Sensation: From Psychophysics to Affective-Motivational Pathways," starts with a controversial and widely reported 2014 report published in the journal *Science* by a team from the University of Virginia and Harvard led by Tim Wilson. Given the abundance of electronic devices and screens constantly surrounding us, their rationale was to see the effect of removing them, and consequently of leaving research subjects alone with their thoughts for a specified time period. One variation of the study introduced the option to self-administer a painful electric shock. Predictably, sensationalist head-lines in the BBC, *Wired,* and *Forbes* claimed that subjects would rather shock themselves to induce pain rather than experience boredom. This chapter uses the Wilson et al. (2014) study as an opportunity to histo-ricize the distinct and successive ways that nervous pathways and af-fective and motor functions of pain have been diagrammed within the body, from the nineteenth- and early-twentieth-century experiments in physiology and psychophysics, which began to offer more defined car-tographies of pain and the nervous system, up to the more recent neuro-scientific model that forms the basis for Wilson's discussion. Inevitably, this features laboratory-based measurement and auto-experimentation on individual subjects but, as the chapter unfolds, we see how physio-logical investigations broadened in scope to diagram pain and sensa-tion within the broader milieu of the organism and its environment. This chapter consists of four sections. The first offers highlights of labo-ratory research that shifted discourses about pain and sensation from a generalized "common sense" *(Gemeingefühl)* of the previous chapter to a more spatially specific cartography of pain and tactile sensitivity in the nineteenth century. It details how such sensations became subject to scientific scrutiny, including E. H. Weber, whose *De Tactu* (On touch) in 1834, and *Der Tastsinn und Gemeingefühl (On the Sense of Touch and Common Sensibility)* in 1846 put the measurement of pain and touch upon a rigorous empirical footing. In turn, Weber influenced Gustav Fechner, whose *Elemente der Psychophysik (Elements of Psychophysics)* of 1860 initiated the whole field of psychophysics, the precise measure-ment of sensation across human subjects. Both Weber and Fechner understood the necessity of objectively measuring thresholds of sen-sation for the individual body as a basis of comparison with others. Utilizing specialized measuring instruments, the resulting Weber–Fechner Law of two-point thresholds helped to produce a cartography of the neurologically "normate" sensory apparatus of the body. Not only did it help dissociate a range of bodily sensations from the larger cate-gory of "touch," but this process of measurement also established what was "normal" and what was "deviant" in terms of capacities, and thresh-olds, of pain and touch sensitivity. Instruments that derived from their experiments were used by later generations of researchers for the measurement of biosocial difference in terms of touch and pain, including the work of the pioneering criminologist Cesare Lombroso in his *L'uomo delinquente* (1878, *Criminal Man*), for example. Second, later theories of pain successively widened the scope of the physiological mechanisms and external influences on the organism, such as Melzack and Wall's cybernetics-influenced gate control theory. Third, we briefly consider the nervous system as a homeostatic system, which finds a historical parallel in explanations of the *milieu intérieur* of the organism, via Claude Bernard's concept of the *milieu intérieur* and Kurt Goldstein's 1934 mixture of neurology and psychology *The Organism*. Fourth, putting the investigations in physiology, psychophysics, and neurology together from the preceding sections, the conclusion argues how pain helps tip the organism as a whole from perception to action, but also operates beyond the organism as a biopsychosocial phenomenon.

"The Oculomotor: Labyrinths, Vestibules, and Chambers" is the third chapter. Experiments by Ernst Mach in the 1870s showed how the vestibular (balance) system of the inner ear was involved in sensations of movement, and how eye movement affected motion sickness. The gross anatomy of the vestibule, semicircular canals, and cochlea of the osseous labyrinth of the internal ear had been known since Galen, and its internal fluid (endolymph) investigated in the seventeenth century by the Italian scientists Valsalva, Contugno, and Scarpa. But it was Joseph Breuer's work on the sensory receptors of the inner ear, in a series of papers between 1873 and 1903, that definitively proved its nonauditory function. In 1873 Breuer delivered a paper to the Imperial Society of Physicians in Vienna, "On the Function of the Semicircular Canals of the Aural Labyrinth," which identified the role of the canals in sensing angular movements of the head as a result of the movement of endolymph within the three planar orientations of the canals. From precursors in amphibians and fish, Breuer inferred a similar mechanism for detecting linear acceleration in the otolith, within the vestibule, which worked in synchrony with the semicircular canals, providing more comprehensive orientation information due to the sensing of movement. Breuer was consequently the first to explain the compensatory nature of eye movement (the way that, when spinning in one direction, the eyes are automatically directed in the opposite direction) and its direct relationship to stimulation of the semicircular canals. During this period, Mach was working on a series of related experiments with his famous spinning chair, published as *Grundlinien der Lehre von den Bewegungsempfindungen (Fundamentals of Movement Perception)* in 1875, and Breuer and Mach corresponded over their discoveries, leading to the Mach-Breuer theory of semicircular canal function. Breuer and Mach's work was revisited in mid-twentieth-century astronaut training, in terms of the spatial reference frame of the oculomotor body, and the phenomenon of vestibular-ocular reflex (VOR) in the understanding of motion sickness. Along with subsystems that deal with balance and orientation, here I offer parallels between spatial motifs of the interior spaces of the body—labyrinths, vestibules, chambers—and those in artefacts and the built environment that contribute to the heightened physicality of an oculomotor subjectivity. After bringing in historical ideas about eye movements at the interface between art history and psychology, oculomotor perception is illustrated through a case study of a built environment, a hypothetical walk-through of an archaeological site. The Roman complex of gymnasium and baths known as a "palaestra" offered visitors an experience of heightened physicality, I argue, not only through the activities on offer, but also through the use of floor mosaics and features designed to draw the moving eye into the wider built environment.

"'The Neuro-motor Unconscious': Étienne-Jules Marey, Eadweard Muybridge, and Motion Capture," the fourth chapter, tells a story of two intertwining characters, the idiosyncratic English artist-cum-scientist Eadweard Muybridge and the rigorous physiologist-cum-engineer Étienne-Jules Marey. Although aspects may be familiar to readers, I retell some biographical details of Muybridge's rather startling photographic motion studies as a way to understand Marey's uptake of similar apparatuses for rather different purposes. Muybridge's iconic multicamera photographs of the horse in motion at Palo Alto were inspirational for many, but frustratingly unscientific for Marey. With the help of engineers and collaborators, Marey innovated upon the photographic apparatuses in order to capture not just the movement of horses' legs but something of the biomechanical essence of animal movement through what he termed *chronophotographie.* Inventions such as the *fusil photographique,* or "photographic gun," of 1882 captured a series of movements on the same photographic plate, thereby producing a spectral streak of motion more suitable to Bergsonian duration.

According to François Dagognet, Marey wished to capture something of the "neuro-motor unconscious" through a variety of equipment and media, to produce representations "made of rhythms, muffled pulses and fluxes traversing the corporeal machine . . . in short, the automatic writing of nature itself" (1992, 232). The writing of natural movement would initially be inscribed by means of instruments directly attached to animals and men in his laboratory, and then more distanced chrono-photographic means, including a whole purpose-built physiological field station in a park in Paris. Marey's career therefore had its own trajectory, from the physiological laboratory that he considered redolent of the blood and death of his predecessors' routine method of vivisection, to extramural observations of animals "in the wild" in parks, fields, and open spaces. Of course, Marey's instruments and chronophotographic techniques were being developed at a time when internal physiological phenomena of movement, rhythm, and pulse were increasingly subject to outward measure and optical representation, where the tendency was "to regard the animal body itself as a matrix of surfaces," according to historian of science Robert Brain (2008). Marey was able to fold the different means of measuring and capturing motion, from the sphygmo-graph, which directly measured pulse in 1861, for example, to a series of cameras with rotating discs, from the handheld *fusil photographique* to a large camera obscura on rails at his physiology station. Throughout his career, such methodological innovations were united by what he called the "graphic method," ways of rendering the invisible aspects of motion more visible. This was in continuation with other experiments with inscription in medicine, physiology, and the arts, particularly between 1847 and 1930, including Ludwig's kymograph (1847), Helmholtz's myo-graph (1849), and Karl von Vierordt's sphygmograph (1854). However, Marey's work was not only wide-ranging in terms of subject matter and technique, it also had practical application, producing measuring equipment early in his career for the circulatory system, for example, and later applying his investigation of animal and human locomotion to methods of gymnastic and military training, as he reports in a series of publications and reports, not only the usual academic monographs (e.g., *La Machine Animale,* 1873; *Le Mouvement,* 1894) but also numerous public lectures and essays in popular science publications such as *La Nature.* With the help of engineer Charles Fremont, his chronophoto-graphic techniques were also used to document the skills of artisans, the better to improve labor practices and to streamline automated assembly lines as a precursor to the Taylorism that would sweep Europe and America around 1913. Such innovations in graphical forms of the observation of workers is central to the following chapter on "fatigue."

Hence, "Fatigue: Jules Amar, Angelo Mosso, and Physiological Observations of Industrial Labor, 1891–1947" is the fifth chapter. In the summer of 1915, the foremost physiologist in Britain, based at the University of Oxford, mounted his bicycle and traveled the sixty miles from Oxford to Birmingham in order to spend three months in the VickersMaxim munitions factory. The physiologist, the very same Charles Sherrington who featured in chapter 1, was conducting research on industrial fatigue, would be elected as the first chairman of the Industrial Fatigue Research Board (IFRB) in 1919, and was the lead author of its first ever report in 1920. The remit of the IFRB, which became the Industrial Health Research Board (IHRB) in 1927, was "having regard both to industrial efficiency and to the preservation of health among the workers." Opening with Sherrington's visit, the chapter shows how the period 1870–1945 was reliably productive in fostering ideas about observing, measuring, and then representing repetitive human movements. This is partly the story of curves: the "work curve" (*Die Arbeitscurve*, 1902) of Emil Kraepelin, and the "curve of fatigue" of Hugo Kronecker (1873), replicated and elaborated upon by Angelo Mosso in *La Fatica* (1891). But it was Helmholtz's famous experiments tracing the movement curves of electrically stimulated frogs' legs that inspired these scientists in the first place. Helmholtz's ideas on the transfer and conservation of energy were also incredibly influential, especially his paper "On the Interaction of Natural Forces" (1854). There, Helmholtz coined the term *Arbeitskraft* (labor power) for forces that operate across organic beings (animals and humans, with their chemical processes), and machines, with their physical forces. In France, meanwhile, Jules Amar was a pupil of Chauveau, the head of the Marey Institute after its eponymous founder died. His *Le Moteur Humain* in 1914 compiled extensive observations and data on the forces, movements, and thermodynamic processes involved in labor, including the physiological effects of fatigue on workers. It was translated into English only after the war, in 1920, as *The Human Motor: or, The Scientific Foundations of Labour and Industry*. Amar referenced Marey extensively, as well as Helmholtz, and this territory would be revisited and developed by Georges Canguilhem in his 1947 lecture

"Machine and Organism." In the United States, research on labor was conducted by Frederick Winslow Taylor in 1911, with his emphasis on the stopwatch, and Frank and Lillian Gilbreth's "micromotion" studies from 1907 onward. They stressed efficiency in movement through systematic observation in the laboratory, starting with bricklayers and then inevitably, in World War I, munitions manufacture, and developed their own measuring devices such as the chronophotographically inspired "penetrating screen." Between the United States and Britain there were divergent outcomes. In Britain, the IHRB reached its peak of efficacy through the 1937 Factories Act, which ruled on acceptable working hours, ambient conditions, and rest breaks. The research of Taylor and the Gilbreths in the United States, by contrast, concentrated on maximizing factory efficiency, while the Harvard Fatigue Laboratory (1927–47) went on to track the physiology of the body in various environmental extremes. Building upon Marey's legacy of chronophotographic work, this chapter also pursues the "extra-mural" application of scientific observation, this time into industrial contexts, a unique period in the swirling accumulation of ideas around economy, society, and human physiology in industrial settings.

Finally, chapter 6 is entitled "Motricity: Merleau-Ponty, 'Motor Habit,' and the Neurology of 'Abstract' and 'Concrete' Movement." Some of the empirical findings discussed in earlier chapters were taken up by a twentieth-century phenomenologist with a strong interest in neurological and psychological case studies, Maurice Merleau-Ponty. In part 1 of *Phénoménologie de la perception* (1945), he outlines the limitations of mechanistic physiology and classical psychology as a means of investigation into the lived spatiality of *le corps propre* (one's own body). There is a prereflective character to this lived spatiality, which will return us to the earlier physiological definitions from chapter 1 of proprioception from Sherrington (i.e., *proprio*-ception meaning the perception of one's own body). Aware of the collaboration between psychologist Aldhémar Gelb and neurologist and psychiatrist Kurt Goldstein, and their string of papers in the aftermath of World War I, Merleau-Ponty focusses on Schneider, one of their celebrated case studies, a man who suffered extensive brain injury. Despite the outward appearance of normal functioning, Schneider's unusual inability to grasp perceptual situations as a whole also affects his ability to accomplish particular movements and motor tasks, and for Merleau-Ponty "clearly shows the fundamental relations between the body and space" (1992, 103). For him, therefore, the Schneider case underlines the centrality of movement *(motricité)* in this lived spatiality of the body. My final chapter therefore elaborates on what Merleau-Ponty intends by his reactivation of the term "motricity" from the physiology literature, and follows this through to its more phenomenological conclusions, involving his unique formulation of the significance of movement within intentionality—that is, "motor intentionality." This discussion will lead us back to one of the central concepts that emerges in *Phenomenology of Perception,* namely the "body schema" *(le schema corporel),* which was itself adapted from his reading of Head and Holmes's neurology paper of 1911. The chapter explains the place of movement, and more particularly motricity, within the body schema, and therefore entails a specifically "motor intentionality," as Donald Landes puts it in his recent translation of Merleau-Ponty's *Phenomenology of Perception* (2012, 112ff). Merleau-Ponty focuses on the centrality of movement to embodied consciousness in the third chapter of his book, at one point deeming *motricité* as "not the handmaid of consciousness" but, rather, the "motor grasping of a motor significance" (1992, 143). This motor significance emerges whenever we learn a new dance move, say, or have to adjust our movements when working at a factory production line. Schneider, recuperating after his war injury, was reduced to doing routine tasks on such a production line, yet any ability to improvise movements or deviate from previous motoric habits was entirely missing for him. Schneider's story therefore reveals what happens in the breakdown of motor signification, his inability to transpose routine movements across into new contexts or situations. While the neurological case study of Schneider has been examined fairly extensively in the literature, for my larger project here Merleau-Ponty's formulation of motor intentionality adds a further theoretical strand, bringing together the rich history of the scientific observation of movement, fatigue, orientation and suchlike into an analysis of bodily situation, action, and perception. This chapter works as the culmination of this book's project, therefore, bringing together the fascinating coproductive relationship between the movement of one's own body, the physiological science of the observation and measurement of sensation and movement, and certain contemporaneous explorations in industrial motion studies and arts practice that captured, traced, and then represented those bodily movements.

CHAPTER 1

The "Muscle Sense" and the Motor Cortex

A Cartography of Bodily Interiority

From the story of rehabilitation, neuroprosthetic developments, and the mapping of fine-grained sensorimotor control that features in the introduction, this chapter doubles back to a series of discussions in the second half of the nineteenth century, primarily in the area of physiology and experimental psychology, on the existence and nature of indistinct and sometimes vague sensations arising from the bodily interior. Although such sensations are familiar to our experience, manifesting in various forms such as fatigue, pain, or muscular resistance, for centuries they were bundled together as forms of "inner touch," as Heller-Roazen (2007) shows. A textbook from 1919, Ward's *Psychological Principles*, identified a "laxity in the use of terms" dealing with sensations felt as arising within the bodily interior, whereby some psychologists had erroneously identified "pleasure and pain as sensations," for example (Ward 1919, 44). Luigi Luciani, in a substantial chapter entitled "Sensibility of the Internal Organs" of his five-volume *Human Physiology* (1917), similarly acknowledged the complexity of labelling such sensations, emphasizing how this problem extended to the laboratory: "Many of the bodily feelings thus classified escape physiological analysis owing to their vague and obscure character" (Luciani 1917, 59). However, Luciani attempted to impose some historical order upon his nineteenth-century forebears in physiological science, as I do here, including discussion of the general or common sensations such as *Gemeingefühl* (common sensibility, or coenesthesia) in Henle (1840) and Weber (1851), or Wundt's *Gemeinempfindung* (common sensations, 1874), referred to by the English psychologist E. B. Titchener's translation as "organic sensations" (1908), for example.

How these sensations became more distinct and consequently measurable through medicalized terminology and the physiological laboratory between 1851 and 1906 is the starting point of this chapter. It

involves dialogues between German and British scientists on the sensations of the moving body, encompassing physiology, neurology, and early
neurosurgical experiments, which together helped pioneer a distinct
"physiological psychology" approach following in the wake of Wundt in
Germany (see, e.g., Sully 1876). George's paper "Der Muskelsinn," for example, argued for a distinct "muscle sense" (George 1870), promptly abstracted in English in *The Journal of Psychological Medicine* (Lincoln and
Webber 1871). From its German roots, the term was investigated more
fully in England, but remained relatively short-lived and often controversial. British physiologist Henry Charlton Bastian devoted a chapter-
long appendix to *The Brain as an Organ of Mind* (Bastian 1880, 691ff)
on "Views Concerning the Existence and Nature of a Muscular Sense,"
where he evaluated evidence derived from the work of his contemporaries, demonstrating how quite active this was as an area of research.
Bastian was subsequently "asked by the Council of the Neurological Society in London to open a discussion on the 'muscle sense,' its nature
and cortical localization" at their 1887 meeting, according to Granit
(1973, 1). There, and in the monograph-length paper "The 'Muscular
Sense'; Its Nature and Cortical Localisation" that resulted in the journal *Brain,* Bastian recognized that interior sensations derived not only
from the muscle tissue but also collectively from tendons, joints, and the
skin, and so introduced a "sense of movement" he termed "Kinæsthesis"
(1887, 5ff) as an alternative to a strictly muscular sense. Likewise, the
neurophysiologist Charles Scott Sherrington's term "proprio-ception"
(1906, 1907) was based on years of laboratory observations on postural
reflexes and his hypothesis of how systems of reflexes interact and inte-
grate within the larger organism. The first half of this chapter therefore
shows how more precise physiologically derived terminology emerges
out of the broader concepts of "coenesthesia," the common or general
senses, the so-called "muscular sense" in the second half of the nineteenth century, and what Titchener (1908) and Elsie Murray (1909) summarize as "organic sensation" in the first part of the twentieth, based on
factors such as the refinement of laboratory techniques that could tease
out and observe sensory and motor nerves from bundles of muscle tissue, and greater fidelity in mapping nerve pathways and cortical areas.

   Around this same period in Germany, the young physicians Gustav
Fritsch and Eduard Hitzig were applying electrical stimulation with
galvanic current to dogs' brains at the Berlin Physiological Institute in

order to map cortical areas associated with movement, leading to their identification of a "motor cortex" (Fritsch and Hitzig 1870). Strangely, it was in that exact year of 1870 that John Hughlings Jackson, the so-called father of English neurology (Gross 2007, 327), independently discovered the motor cortex, and his findings were backed up by David Ferrier in a series of experiments on monkey brains from 1873. Ferrier discussed the "motor centres" of the cortex in terms of voluntary movement and paralysis in a section of his *Functions of the Brain* (Ferrier 1876, 199ff). Ferrier's experiments, along with early neurosurgical work by Victor Horsley, encouraged the American neurosurgeon Harvey Cushing to conduct experiments to map the motor parts of the human brain between 1902 and 1912. For the purposes of this chapter, however, the link between the neurophysiological discoveries of the organs and sensations of movement, on the one hand, and the budding neurosurgical experiments with cortical mapping, on the other, come together in Sherrington's work. For Sherrington directly inspired one of his students in turn, the American-born neurosurgeon Wilder Penfield, to use direct brain stimulation to map precise cortical areas associated with each of the senses. It was through building upon the work of Ferrier, Sherrington, and a surgical apprenticeship with Cushing in particular, that led Penfield to create maps of the sensory and motor cortices of the brain depicted as a "homunculus" in the 1930s. The homunculus was first realized in a collaboration with Edwin Boldrey and the medical artist Hortense Cantlie, who first produced the world-famous and widely reproduced illustration, divided in fact into a "sensory homunculus" and a "motor homunculus" in a paper by Penfield and Boldrey (1937), "Somatic Motor and Sensory Representation in the Cerebral Cortex of Man as Studied by Electrical Stimulation." The idea and the illustration itself were updated in a later monograph coauthored with Rasmussen, *The Cerebral Cortex of Man* (Penfield and Rasmussen 1950). Of course, variations of the famous "homunculus" illustration are reproduced in innumerable psychology textbooks to this day as a map of body sensation and motor control. Although no longer taken seriously by researchers, the illustration still inspires other neurophysiological mapping projects, surprisingly. For example, Mazzola et al. (2012) revisited Penfield's surgical techniques of direct cortical stimulation to assess pain perception. But far more significantly for our purposes, from a historiographic perspective Paula Di Noto et al. (2013) found that

Penfield had mapped both female and male patients, but for some reason the female data were not incorporated into the model. They revisited Penfield's archives in order to reclaim the data from female subjects to produce an alternative to the androcentric homunculus they call a "*her*munculus." The story of the formation of the homunculus, divided into a "sensory homunculus" and a "motor homunculus," and ultimately the reclaiming of the "hermunculus," derives from photographs of pioneering open-head brain surgery and ink illustrations prepared for publication by Penfield, archived at the Montreal Neurological Institute (MNI). Archival research I conducted at the MNI aids this narrative, where I was able to observe the original images and markup of these illustrations, filling out the context a little. Quite apart from the infamous homunculi drawings, Penfield and his collaborators laid the foundations for what is now called the primary somatosensory cortex ("S1").

This chapter consists of two parts, proceeding from the muscles as the localized site of sensations of movement and position to early discoveries in the motor and somatosensory cortex, which paves the way for more precise cortical mapping of movement and sensation through neurosurgery. The first part, "The Muscles as an 'Organ of Sense,'" starts with Aristotle's concept of the faculty of bodily sensation, or *aisthêsis,* and those sensations felt within the bodily interior such as movement, pain, fatigue, and hunger. But this is to set the scene for long-running nineteenth-century debates among physiology, psychophysics, and psychology on the existence or otherwise of a "muscle sense" (from the German *Muskelsinn*) or "muscular sense." The recognition of a distinct "muscle sense" occurred in Anglophone physiology initially through Charles Bell (1833), but sparked discussion of a series of related and increasingly medicalized terms, including the concepts of "kinaesthesia" from Henry Charlton Bastian (1880, 1887), and "proprio-ception" from Charles Sherrington (1906, 1907). Sherrington's extensive research on integrating reflexes for the sake of ordering an organism's orientation and position is examined in some detail. The second part, "The Motor Cortex: Penfield's Cartography and Cantlie's Creature," retells the story of the formation of a rather unique science–arts collaboration, their sensory and motor homunculus. This offers an opportunity to reexamine the import of Penfield's greater mapping project of the motor and somatosensory cortices, and note its shortfalls in excluding female subjects and cherry-picking data to support a particularly normative and static diagram of cortical representation—that is, a physicalist snapshot of a relationship between the mind and the body that could be otherwise.

The Muscles as an "Organ of Sense"

Before examining how more medicalized forms of inner touch and the muscle sense emerged through nineteenth-century physiological investigations, something of the prescientific background and assumptions around muscles and movement are briefly outlined. For this we turn briefly to Aristotle to establish the pervasiveness of a generalized "inner touch" faculty from their inquiries well into the Early Modern period (for extended discussion of this history, see, e.g., Kemp and Fletcher 1993; Heller-Roazen 2007; Paterson 2007, 2012, 2013).

The feeling of muscular effort, or of fatigue localized in a particular body region, is familiar to anyone who walks any distance, especially if it involves a steep incline. Common explanations for these sensations include a buildup of lactic acid in the muscles, possibly combined with an understanding that, in this instance, our movements have departed from their usual habitual range. *Problemata,* a text attributed to Aristotle but widely believed to involve the work of the Peripatetic School after his death, consists of a connected series of brief hypothetical "problems" in question-and-answer form.[1] In book 5, "Fatigue," the author considers this exact scenario of walking uphill in order to introspect on the common bodily experiences of fatigue, ache, and pain in moving bodies. This is possibly the first time that the interiority of the body has been subject to such philosophical inspection. The relevant sequence of problems starts like this:

> Why is it that in ascending a slope our knees feel the strain, and in descending the thighs? Is it because when we ascend we throw the body upwards and the jerk of the body from the knees is considerable, and so we feel the strain in the knees? But in going downhill, because the weight is carried by the legs, we are supported by our thighs, and so they feel the strain. (882b25–31)

In the proceeding problems, he revisits and refines his questioning, seeking material answers based on this perceived localization of sensation in the limbs. He soon equates the repetitive movement of limbs to other, nonhuman objects that have the capacity for movement in the physical world, arguing that the greatest pressure is always felt at the center of any moving object: "Now the thigh is of this nature and so it is in the middle of it that we feel the strain most" (882b37). Next, he considers the spatial relationships between the fatigued limbs and other parts of the body. Since the moving body generates heat, he wonders about the effects upon contiguous areas, whether the heated body intensifies the felt sensations in the upper leg as opposed to the lower, and consequently makes them expand or contract more readily:

> Why is it that the thighs feel fatigue more than the legs? Is it because they are nearer to the part of the body which contains the excrement? . . . If one feels fatigue when there is no excrement in the body, even so it is the thighs and loins which suffer more than the other parts. (883b14–21)

Because of the short question-and-answer format, the introspective line of questioning rapidly diffuses into other considerations, such as the physics of running as opposed to walking, and so on. Written at a time with limited understanding of the physiology of muscles and nerves, Aristotle seeks an explanation for sensations of fatigue or ache because of certain spatial relationships with respect to the human body and its environment. As Sheets-Johnstone puts it, these two explanations "hinge on *spatial relationships,* spatial relationships obtaining between centers and peripheries in the case of stresses and strains, and spatial-elemental relationships obtaining between movement and heat, and between excrement and the proximity of thighs in the case of fatigue." The sensation of strain and fatigue only occurs, she explains, because "certain spatial or spatial-elemental relationships obtain with respect to the world generally and to the human body in particular" (1992, 136–37, emphasis in original). In other words, sensations of bodily interiority are somehow aligned and interpreted according to a spatial schema with a center and a periphery, and for Aristotle are hypothesized in baldly material terms of cause and effect between contiguous body parts. Although a lesser-known Aristotelian text, this sequence of problems remains notable in that it starts from an easily recognized perspective, the phenomenology of felt aches and fatigue in the bodily interior that most people experience in some form. As Sheets-Johnstone summarizes, "What has changed in 2,300 years is not the body and our experience of it, but ways in which we conceive it" (137). Elsewhere, in *De Anima (On the Soul),* Aristotle discusses *aisthêsis,* the faculty of perception, in terms we currently understand as both "sensation" and "perception." He praises the nobility of sight but acknowledges that all animals, including the lowly ones, share the capacity for touch (*De Anima* 413b4–7) at least, and that touch and other senses interact in the perception of objects. Various properties of objects such as size, shape, number, and movement are perceived incidentally by a number of senses (425a23) working together, so-called common sensibles. However, rather than positing a distinct, unifying "common sense" that stands outside the five sensory modalities, sometimes *aisthêsis* is referred to as straightforward sensation, and sometimes as a faculty of perception that involves judgment on behalf of the perceiver. For Aristotle in *De Anima,* says Hamlyn, "the faculty of sense-perception is that faculty by means of which we are able to characterize or identify things as a result of the use of our senses" (1959, 6). In *De Sensu et Sensibilibus,* usually translated as *Sense and Sensibilia* (e.g., Beare's translation in Aristotle 1984), Aristotle infers the existence of a general or total sense *(aisthêtikon pantôn),* a faculty able to correlate percepts from different organs of the body, which has the function of a more general sensory capacity: "one faculty in the soul with which the latter perceives all its percepts, though it perceives each different genus of sensibles through a different organ" (*De Sensu* 449a). This is a very different *aisthêsis* from Plato's discussion in *Theaetetus,* for example. Aristotle and the root concept of *aisthêsis* forms a productive wellspring for reconsidering the type and nature of experiences that arise from perception through the body (see, e.g., Paterson 2012, 2013).

The trace of these Aristotelian ideas of the bodily interior was felt throughout the Middle Ages. It is consistent with the pervasive idea of "a kind of inner touch, by which we are able to grasp ourselves," as Heller-Roazen (2007, 241) declares. This long trajectory originates in classical scholarship but is something the Abbé de Lignac terms "the sense of the coexistence of the body" in his *Elements of Metaphysics Drawn from Inner Experience* of 1752, and a series of sensations that Turgot in his *Encylopédie* entry of 1755–56, acknowledging Lignac, wished to place under a special class of sensation beyond the assumed five senses. Turgot acknowledged other, "so to speak more penetrating senses, which can be referred to the inside of our body," which he classified "under the name of *inner touch,* or the sixth sense" (in Heller-Roazen 2007, 242). This special sense included the kinds of aches, pains, hunger, thirst, even intestinal murmurs and "shudders" that may not be precisely located, or always acutely defined against the background of bodily sensation, "the whole multitude of confused sensations that never leave us; the sensations that, so to speak, circumscribe our body, rendering it always present" (Turgot, in Heller-Roazen 2007, 242). Toward the end of that century, the German physician Hübner differentiated between external perception, the idea of inner touch, and something like a general feeling of the organism's own sensation. Theoretical speculation about the sensations that arise in the flexing of muscle tissue were already underway in the first third of the nineteenth century. As early as 1820, Thomas Brown had identified sensations deriving from "an awareness of muscular contraction" in his *Lectures on the Philosophy of the Human Mind,* according to Jones (1972, 299). In Brown's lectures 20 to 22 on touch, for example, the sensations particular to a muscle sense are recognizably distinct and worthy of attention:

> The feeling of resistance is . . . to be ascribed not to our *organ of touch,* but to our *muscular frame, . . .* as forming a distinct *organ of sense. . . .* The sensations of this class, are . . . commonly, so *obscure,* as to be scarcely heeded . . . but there is no contraction, even of a single muscle, which is not attended with some faint degree of sensation, that distinguishes it from the contraction of *other muscles* or from *other degrees of contraction* of the same muscle. (Brown 1820, 496, emphasis in original)

His focus on a form of sensation that is distinctly muscular in origin, yet actually imprecise and vague, is retained, so that shortly afterward he states: "Each motion of the . . . limb, whether produced by *one* or *more* of the . . . muscles, is accompanied with a *certain feeling . . .* which we distinguish from every *other feeling* accompanying every other *quantity of contraction*" (497, emphasis in original). In other words, a more general awareness of bodily position is maintained through more localized and distinct muscular tensions and contractions. In line with his forebears, this is interpreted as a form of bodily touch that extends beyond the cutaneous and proposes the whole body as an organ of sense, wherein "our muscular frame is not merely a part of the living *machinery* of *motion,* but is also *truly an organ of sense*" (501, emphasis in original). In this way, Brown, and later George, Wundt, and their contemporaries, could be considered as rearticulating the ongoing inner touch of *aisthêsis,* while focusing on its manifestations through muscle fibers and the muscular frame. Charles Bell, the Scottish anatomist and surgeon, also claimed a specific "muscle sense" in *The Hand: Its Mechanism and Vital Endowments as Evincing Design* (1833), and Sir William Hamilton, in one of his supplementary essays for his 1846 edited collection of the works of philosopher Thomas Reid, discussed the "feelings of movement" of a limb, according to Bastian (1869, 461).

Against this diffuseness of inner touch, a more refined answer that distinguishes the parameters of a "muscular sense" starts with Charles Bell's 1823 essay "On the Motions of the Eye, in Illustration of the Uses of the Muscles and the Orbit," followed by his 1826 essay "On the Nervous Circle Which Connects the Voluntary Muscles with the Brain," and developed further in his 1833 monograph *The Hand,* where he deliberates on the interaction between touch and movement. In his 1823 paper to the Royal Society on eye motions, he reports on his anatomical discoveries of distinct, specialized sensory and motor nerves that led him to investigate the mechanisms of the nervous system that governed and regulated muscular activity. In it he suggests that a muscle sense is involved in the maintenance of balance: "Let us consider how minute and delicate the sense of muscular motion is by which we balance the body, and by which we judge of the position of the limbs, whether during activity or rest" (1823, 181). Wade (2003) argues that in his 1823 paper Bell provided evidence for a muscle sense, but his 1826 paper carried more weight, including more evidence from his anatomical dissections in combination with his physiological speculations. A section that dealt with our capacity for discriminating different weights in the hand offers a phenomenological component to his physiological hypothesis, argues Wade, who also connects the technique of comparing weights to what E. H. Weber will do a few years later in his psychophysics laboratory (Wade 2003, 185). In the monograph *The Hand,* Bell again proffers a distinct muscle sense, this time in terms of a "consciousness of muscular exertion" akin to a sixth sense (1833, 195). Given this customary sense of muscular coordination, an awareness that heightens with exertions and spasms or even the estimation of weights through the use of our hands, Bell then states his goal: *"I shall first enquire, if it be necessary to the governance of the muscular frame, that there be a consciousness of the state or degree of action of the muscles."* This can be asked since "we are sensible of the most minute changes of muscular exertion, by which we know the position of the body and limbs, when there is no other means of knowledge open to us" (1826, 167, emphasis in original). In passing he offers the example of a ropedancer or a blind man balancing their body, finding an explanation in neuroanatomical terms. Referring to prior discussion of the interactions of sensory and motor nerves, Bell summarizes in deceptively straightforward language: "Between the brain and the muscles there is a circle of nerves; one nerve conveys the influence from the brain to the muscle, another gives the sense of the condition of the muscle to the brain" (170). The hypothesis of a bidirectional nervous mechanism is novel and influential, and in his later book he refines the definition: "There is a nerve of sensibility to convey a sensation of the condition of the muscles to the sensorium, as well as a nerve of motion for conveying the mandate of the will to the muscles" (1833, 196). Conceptualized therefore as a circuit of sense in the nervous system within a muscular-skeletal framework, it could be said that the body is starting to become sensorimotor.

On the one hand, the utility of providing musculoskeletal positional feedback from the muscles to the sensorium forms part of an intuitively recognizable background of embodied experience. As such, Bell's explanation might be considered a more neurologically specialized explanation for those earlier functions of *aisthêsis*. Awareness of a musculoskeletal frame of perception becomes foregrounded in nonvisual experiences, and becomes more prominent for congenitally blind subjects who must determine bodily position in space through nonvisual means, especially their "muscle sense," as Hocheisen (1893) will discuss (see below). But Bell also notices that, without touching anything, a blind person may continually sustain and adjust their upright posture: "It is obvious that he has a sense by which he knows the inclination of his body, and that he has a ready aptitude to adjust it, and to correct any deviation from the perpendicular," and given their lack of vision the only source of knowledge is literally through the body, "a sense of the degree of exertion in his [*sic*] muscular frame" (1833, 198). So far, Bell's discussion of a hypothetical blind man perceiving their own musculoskeletal position in space is an anticipation of what Bastian will term "kinaesthesia" in 1887 and Sherrington will later term the "propriocep-tive" sense in 1906. This awareness of one's own body and the muscular sense is particularly acute when anticipating the possibility of future movement, he recognizes: "We could not command our muscles in standing, far less in walking, leaping or running, had we not a percep-tion of the condition of the muscles previous to the exercise of the will" (200), says Bell. The act of touching similarly involves movement or the anticipation of movement through the prehensile organ of the hand, since it is the "combined perception" (205) of touch with movement of the hands, arms, and fingers that are necessary to embrace objects in the active register necessary for an engaged sense of touch. Here Bell confirms the role of muscular awareness in touching, akin to what E. H. Weber will characterize as "active touch" in his near-contemporaneous *Der Tastsinn* of 1834:

> So it is affirmed by physiologists . . . that the sense of touch differs from the other senses by this circumstance—that an effort is propa-gated towards it, as well as a sensation received from it. This confu-sion obviously arises from considering the muscular agency, which is directed by the will during the exercise of touch, as belonging to the nerve of touch properly. We proceed to show how the sense of motion and that of touch are necessarily combined. (197)

Elsewhere, Bell writes of the positively felt sensations that accompany the mastery and refinement of movement. The pleasure that arises from muscular activity derives in part from a gratification that accompanies any refinement or active shaping of activity. Having attempted to iden-tify the nervous circuits that account for mixed sensations of movement and touch, he also discusses the kinds of pleasures and enjoyments that result in a section of *The Hand* entitled "The Pleasures Arising from the Muscular Sense":

> The exercise of the muscular frame is the source of some of our chief enjoyments. The beautiful condition of both body and mind shall result from muscular exertion and the alternations of activity and bodily repose. . . . This activity is followed by weariness and a desire for rest, and although unattended by any describable pleasure or

local sensation, there is diffused throughout every part of the frame after fatigue a feeling almost voluptuous. (205–6)

As evidenced here, toward the end of the book Bell starts to depart from a strictly scientific framework for understanding movement and touch, and exhibits his own familiarity with muscular sensations. He furthers this departure by loosely remarking on the aesthetic and even joyful qualities of touch and movement, harking back to classical cultures: "It would appear that in modern times we know comparatively little of the pleasures arising from motion," he says, whereas "the Greeks, and even the Romans, studied elegance of attitude and of movement" (206). Of course, qualities of the harmony of movement, of control of gait and posture, are present in some form in all cultures, but here Bell briefly opens a pathway between science and the arts, the physiology of the muscular sense and more embodied, kinesthetic, and empathic concepts within dance and performance (e.g., John Martin's *America Dancing,* 1936) that has intensified in the past decade (e.g., Susan Leigh Foster's work, including *Choreographing Empathy,* 2011). However, after this spate of interest in the science and experience of the muscular sense, it would be decades before the concepts, and then the terminology, of a distinct set of muscular sensations including fatigue would become resurrected by German physiologists. It is in a roundabout way, through translations and reports, that Anglophone physiological science reacquaints itself with Bell's ideas, then.

COMMON SENSATIONS, GENERAL FEELINGS, STRANGE SENSATIONS

Physiological terminology in Germany made a pronounced shift toward the general or "common," through the prefix *Gemein-* (common)—for example, Henle's *Gemeingefühl* (common feeling), or Wundt's *Gemein-empfindung* (common sensation), discussed below. The idea of such a generalized capacity for bodily sensing, with its conceptual roots in *aisthêsis,* would therefore persist well into the nineteenth century, but the French term *cénesthésie,* anglicized to "cenesthesis," or "coanesthesia," by Sir William Hamilton in 1851, retained the actual etymological stem. Francis Schiller in fact argues directly that coenesthesis was specifically coined in German by Reil to combine "*koinos* (common) and *aisthesis* (sensation, feeling) to express the [then] current need to say

Gemeingefühl in Greek" (1984, 496). Maine de Biran, in his "Écrits sur la physiologie" of 1821, was apparently the first francophone writer to import the term from Reil's German, defining *coenesthèse* to those background sensations of "pleasure or pain, which [are] inseparable from everything that lives," but also characterized as the "general feeling, the composed modality of all the vital impressions inherent in every part" of the living organism (Maine de Biran, in Heller-Roazen 2007, 243). Meanwhile, Jacob Henle employed the term "coenaesthesia" in a more specifically medical context in his *Pathological Investigations* (1840) and *Anatomy* (1841), characterizing it as "the tonality of the sensitive nerves" in general, and "the sum, the confused chaos of the sensations which are instantly transmitted to the brain from all parts of the body" (Henle, in Luciani 1917, 59). Tellingly, by the time that Moritz Schiff wrote the lengthy treatment of *cenestesi* in the Italian *Dictionary of Medical Sciences* in 1871, the concept of coenesthesis was somewhat muddied by the presence and proliferation of terms from Germans other than Reil, who had emphasized the general or common aspects of bodily sensation. A decade after Henle's "coanesthesia," then, Ernst Heinrich Weber's laboratory experimentation would yield a comparable term, *Gemeingefühl,* or "common sensibility," in his *Der Tastsinn und Gemeingefühl* (1851), being in his words "the consciousness of our sensory condition" (1978, 213). These and other such formulations reveal the preponderance of conceiving the manifold inner senses as a form of touch. Although the new science of psychology began in earnest with Helmholtz's student Wilhelm Wundt, and Wundt's first laboratory in Leipzig in 1879, E. H. Weber's experimental work in the 1830s had already differentiated *Tastsinn,* or cutaneous touch felt specifically on the skin surface, from a generalized somatic sensation or common sensibility, *Gemeingefühl.* A central element of Weber's *Der Tastsinn und Gemeingefühl* (1851), translated as *The Sense of Touch and Common Sensibility* first in *E. H. Weber: The Sense of Touch* (1978) and then in *E. H. Weber and the Tactile Senses* (1996), was in fact "the ability to perceive our own sensory states," he says (1996, 213), sensations arising through the body including pain, as discussed further in chapter 2. Whether known as Weber's *Gemeingefühl* (common sensibility, 1851), Friedrich Kirchner's *Gefühlssinn* (sense of feeling, 1886), or Wundt's *Gemeinempfindung* (common sensation, 1874), physiological psychology offered a kind of empirical revisitation of Aristotle's *aisthêsis.* For Kirchner's *Katechismus Der Psychologie*

(1883), for example, adapted and translated into English in 1888, the sense of touch broadly followed the Aristotelian categorization in that it was distinguished from the four other senses. The sense of touch was special for Kirchner in that it encompassed "a whole series of different sensations" relating to the body, including pressure, "muscular sensations, sensations of temperature and of pain" (Kirchner 1888, 188). At one point he clarifies exactly this relationship between touch and muscular sensation:

> The sense of touch finds the greatest help in *muscular sensations, i.e.*
> those which correspond to muscular movement. Muscular sensations
> accompany tension and flexing of the limbs, whether the impulse
> proceed from the brain, or, as reflex movement, from the spinal cord,
> whether it be directed to the initiation, furthering, or inhibition of
> a movement, or to the taking up or maintaining of an attitude. (190,
> emphasis in original)

Thus for Kirchner, like Brown decades before him, muscular sensations are distinct from, yet related to, touch. Kirchner only mentions the term *Muskelsinn* twice, and only briefly in the *Katechismus,* but it is clear that muscular sensations are distinct from the kind of cutaneous touch that Weber terms *Tastsinn* (1851), or even generalized bodily feeling *(Gefühlssinn)* as Kirchner himself (1886, 220) termed it elsewhere, or Wundt's common sense *(Gemeinempfindungen),* as Titchener reminds us (1908, 158). To add briefly to the terminological confusion, Wilhelm Volkmann, a philosopher and psychologist with an interest in aesthetics, interpreted Wundt's *Gemeinempfindung* in terms akin to an embodied or somatic consciousness in his *Lehrbuch der Psychologie* of 1856:

> By common sensation *[Gemeinempfindung],* which, because it is not
> sensation, should actually be called common sensibility *[Gemeinge-*
> *fühl],* we understand the overall impression of all simultaneous sensa-
> tions: the somatic consciousness *[das somatische Bewusstsein]* or, as
> it has also been called: the vital conscience *[das vitale Gewissen],* the
> physiological climate. (in Eisler 1904, 284, my translation)

Needless to say, some conceptual confusion reigned across these etymologically related terms. As Murray rather unsympathetically argues,

Weber, Wundt, and Goldscheider were guilty of "lumping organic or visceral sensations with the *Gemeinempfindungen* ('common sensation'), a class which by definition consists in whatever residue of sense material proves resistant to elaboration, analysis, and localization" (1909, 390n1). And, whereas the "common" element across these common sensations for Müller seems to involve a topologically distributed set of sensations and includes the "pain-pressure-temperature equipment common in his belief to the exterior and interior of the body alike" (390n1), the "common" aspect of Weber's *Gemeingefühl* is more epistemologically distinct, excludes externalized sensation, but does potentially include muscle sensation. Before the recognition of, and debate around, the distinctness of muscular sensations in the physiology literature, we briefly clarify these multiple and competing terminological strands that revolve around the "commonness" of sensation.

Rousseau had argued in book 2 of *Emile* (1762) that we are not masters of each sense equally, and that touch is a special sense. Although touch is "spread over the entire surface of our body as a continual guard to warn us of all that can do it damage," argues Rousseau ([1762] 1979, 133), this functionality is not restricted to the skin surface. Friedrich Kirchner quotes this passage in an entry in the *Wörterbuch der Philosophischen Grundbegriffe* (Dictionary of Philosophical Concepts), originally published in 1886 in order to stress the place of touch in the bodily interior: "it also contains a large number of internal organs (joints, muscles, tendons, bones, etc.), it gives us the sensations of pressure, cold, heat, and pain. The sensations of pressure are usually called tactile sensations *[Tastempfindungen]*; the other sensations of the general sense are called common sensations *[Gemeingefühl]*," he writes (Kirchner 1907, 221, my translation). Likewise, Wilhelm Wundt's *Grundriss der Psychologie* of 1896 (soon translated into English as *Outlines of Psychology* in 1897) involves discussion of *Gemeingefühl* at various points throughout the text, including an entire section. Here Wundt begins by pointing out a misunderstanding in terms of the "combination of partial feelings into a composite feeling," or *Gemeingefühl*. Effectively this is a mismatch between an introspective form of "psychology," as he puts it, and the actual bodily "physiology," such that the (subjective) "feeling was not distinguished from its underlying sensations" (1897, 178). This is a problem because *Gemeingefühl* is sometimes simply described as "consciousness of our sensational state," and sometimes more as the "'totality, or unanalyzed chaos of sensations' which come to us from all parts of our body" (178, quote uncited). Although *Gemeingefühl* has a subjectively felt unity or coherence, it nonetheless "consists of a number of partial feelings" that are distinguishable from the usual five senses. Unlike the localization involved in the five senses, then, these "partial feelings" are physiologically more dispersed throughout the body, and their character, or "affective tone," may alter and shift in significance according to the areas and activities of the body at that time. Wundt gives two brief examples that exhibit this felt unity in spite of the dispersed physiology. One is pain, which involves a sudden and strong local sensation but also an accompanying generalized pain feeling. However, as we see in the following chapter, the foundations of pain research from the 1840s onward through Fechner, von Frey, and Wundt himself would establish it as a distinct sensation rather than as a component of *Gemeingefühl*. The second example is more pertinent to muscular sensation, being "the case of certain sensations connected with the regular functioning of the organs, such as the inner tactual sensations accompanying the movements of walking" (178). Reminding us of Aristotle's discussion in *Problemata,* although it may be recognized in its unitary character as touch, or movement, or in fact fatigue, the partial feelings are more particularly localized as sensations in the legs when walking, say, or the arms when swimming:

> Often the relatively greater importance of a single sensation is so slight that the predominating feeling can not be discovered except by directing our attention to our own subjective state. In such a case the concentration of the attention upon it can generally make any partial feeling whatever predominant. (178)

Although the terminology mutated somewhat, the need to pinpoint and scientifically verify these partial feelings remained a concern into the twentieth century. Murray's 1909 essay "Organic Sensation," for example, offered a historical overview of what she collectively describes as the "sensory contributions from the internal tissues" (1909, 400). Reviewing physiological work from the nineteenth-century onward, she remarks upon "the least developed and systematized sphere of our consciousness" in order to "throw new light on processes of localization and attention" (402). Digestive, muscular, and respiratory systems each produce their own sensations, what Ebbinghaus in 1902 terms *eigenartige Empfindungen* (strange sensations). Again, Meumann in a 1907 article "On the Sensibility of the Internal Organs" celebrates a multiplicity of "organic sensations" *(innere Tastempfindingen)* derived from distributed organs and tissues, but actually a direct translation of Meumann's phrase would offer the now familiar "inner touch sensations." The unusual combination of qualitative diversity yet indistinctness of these sensations together obscures "the indefiniteness of localization of the sensations, and their deficiency in correlated visual images by which qualitative isolation might be facilitated," Murray argues (400–401). In other words, these sensations remain vague and unsystematic, and cannot be compared to the kind of clarity that vision enjoys. At the end of one section, Murray summarizes this indistinctness in frustrating terms:

> The fusability, absence of memory images, unanalyzability, lack of cohesiveness with other sensations, unlocalizability, capacity for eluding the attention, and other features ascribed guardedly or confidently in various quarters to our organic experience, demand critical verification. (402)

Meanwhile, James Ward's *Psychological Principles,* for example, mentions "the existence of a background of organic sensations or somatic consciousness," explicitly citing Volkmann's *Lehrbuch* for the latter formulation (1919, 249). Ward had won a fellowship with a dissertation entitled "The Relation of Physiology to Psychology," part of which was published in the first issue of *Mind* in 1876. Bertrand Russell, a coauthor of *Principia Mathematica* (1910–13), and G. E. Moore, author of *Principia Ethica* (1903), were among his students. In Göttingen, he attended lectures by Rudolf Hermann Lotze. Ward makes the connection between the German *Gefühl,* the French *sentiment,* and the English term "feeling," but in trying to pin down this concept further he chooses the "organic sensations" (1919, 41). Ward understands that a feeling sometimes accompanies changes in bodily or conscious states, but is not reducible to that change. He states: "We are aware of a certain change that has occurred either 'in things without or in our thoughts within'" (43), that this change is subjectively for us either painful or pleasurable, and that, as an organism, we continue the activity that will continue the pleasurable feeling or discontinue the painful one. There is a confusion of sensation with feeling, for example, pain. We discuss the demarcation of pain as a distinct and measurable object of study in the following chapter.

A "MUSCULAR SENSE"

If the physiological experimentation by figures such as Weber, Wundt, and Kirchner were indeed guilty of lumping together partial sensations into categories that were general or common, a more sustained experimental attention to the particularity of what is sensed through the muscular tissues becomes of interest in Germany as the *Muskelsinn* (muscle sense) by a Professor George (1870), followed by Martin Bernhardt (1872), Karl Schäfer (1889), Paul Hocheisen (1893) on this muscle sense for the blind, and Alfred Goldscheider (1889) published his "Über den Muskelsinn und die Theorie der Ataxie" (On the muscle sense and the theory of ataxia) (see, e.g., Baldwin 1905, 1173ff for an overview of contributions in this area). George, a German physiologist apparently without a given name, published his "Der Muskelsinn" in *Archiv für Anatomie, Physiologie und Wissenschaftliche Medicin* in 1870, and a year later his paper was reported and discussed in the anglophone scientific community in *The Journal of Psychological Medicine* under the heading "The Muscular Sense *(Muskelsinn)*" (Lincoln and Webber 1871, 396–98). The particular tensions and contractions that make up the *Muskelsinn* become foregrounded in the absence of other sensory stimuli, especially sight. Attention to the musculoskeletal frame of the body, albeit perhaps with the potential for amplified acoustic acuity and resonances, is heightened in the case of blindness, as Hocheisen's essay "Über den Muskelsinn bei Blinden" (1893, On the muscle sense of the blind) evidences. Hocheisen's paper was summarized in the British journal *Mind* of that year:

> The perception of passive movements—the perception of the "muscle sense" in general—are more than normally acute in the case of practiced blind subjects. The causes of this are mental: attention and practice in the interpretation of sensations of slight intensity. The movement-sensibility of children is better than that of adults. The differences of "right" and "left" have but little to do with acuteness of movement-discrimination. The cutaneous "sense of place" is not always probably more than normally acute in the blind: where it is so, the fact is referable to practice. (1893, 553–4)

To some extent this was coextensive with, and qualified by, pioneering physiological discoveries on the nerves present within muscle tissue. As Edwin Boring shows, Kölliker in 1862 first identified what is known as the "muscle-spindle," although he misdescribed them as "nerve buds"—that is, a stage in the development of muscle. Kühne added to the description a year later, in 1863, and correctly named them "muscle spindles," and "suggested cautiously that they might be receptors for the muscle sense," says Boring (1942, 530). Carl Sachs, in his essay "Physiological and Anatomical Examinations of the Sensible Nerves of the Muscles" (Sachs 1874), a few years after George's "Der Muskelsinn" article and in the same journal, saw how the sensory nerves terminated in the spindles. Sachs also dismissed earlier speculation, including that of Helmholtz, that the effort of producing a movement is felt through a combination of the state of contraction of the muscle, the altered position of the limbs, and the tension of different parts of the skin. This would imply the existence of special centripetal nerve fibers in the muscles. In dissections of frogs, Sachs separated out the sciatic nerve, and teased out isolated "nerve twigs," as Coats reported (1875, 385). With the aid of a microscope, some of the nerve twigs were irritated, and of those irritated, some caused involuntary leg movement and some did not. Under the microscope it was found that, in the case of irritated nerves that did not cause muscular contraction, the nerves retained their connection with the centers, and so must be sensory rather than motor. There was now categorical proof that muscle tissue included sensory nerves. While the motor nerves are distributed in bundles, there are comparatively fewer medullated—that is myelinated, or sheathed—sensory fibers. An L. Kerschner (not to be confused with Friedrich Kirchner), who had published his research on *Muskelspindeln* in 1888, argued specifically for this "sensory nature of the muscle-spindle, and suggested that their function might be connected with the sense of position," according to Batten (1897, 175). In the light of such neurophysiological discoveries, the sensory component of muscular movement now had a more scientific basis.

Not everyone believed in the existence of a dedicated and distinct muscle sense, necessarily. Although the portmanteau *Muskelsinn* gained some currency from the 1870s both in Germany and, through its echo "muscular sense," in Britain, there was dissent among fellow physiologists such as David Ferrier (1874) and George Lewes (1878). In the very first issue of the journal *Brain* in 1878, for example, George Henry Lewes published an overview of this area of research, starting from Krause's then-recent discovery of nerve fibrils within muscle tissue in 1876. Lewes's essay, "Motor-Feelings and the Muscular Sense," commences with the acknowledgment of the familiarity of the sensations that arise in the movement of limbs:

> The fact that we have feelings accompanying the movements of our limbs and contractions of muscles, and that such feelings are specifi-cally distinct, is admitted on all hands; the assignment of these feel-ings to the motor apparatus, and their subsumption under the rubric of a Muscular Sense co-ordinate with the other Senses, is one of the vexed questions. (Lewes 1878, 14)

William James, in the second volume of *Principles of Psychology,* also discussed the role of muscular contraction and the movement of joints in the perception of space. In his typical style, touching upon figures we have encountered such as Brown, Goldscheider, and Lewes along the way, James ends up unequivocally dismissing the term: "This word is used with extreme vagueness to cover all resident sensations, whether of motion or position, in our members, and even to designate the sup-posed feeling of efferent discharge from the brain" (James 1890, 197). The discharge from brain to muscle of a limb was based on an assumption from Bain, via Müller, of "a sensation of energy exerted by the outgoing stream" as James put it earlier in *The Feeling of Effort* (1880, 25). A couple of decades later, the otologist Pierre Bonnier in his book *L'orientation* of 1900 agreed with James that there was no experimental justification for a distinct muscle sense. According to a report in English of this pub-lication in the *British Medical Journal* of that year, Bonnier "objects to the generally-accepted term 'muscular sense,' and does not admit that either physiology or pathology affords any justification for its adoption" (*BMJ* 1900, 1393). Reviewing Wundt's position, who thinks erroneously like Müller that there is "a consciousness of nerve discharge which puts muscles into operation, to that which attributes to the muscle itself the power of exciting sensations, giving cognizance of its position at any moment," Bonnier rejects the need to invoke a novel sense for this pur-pose, claiming instead that these phenomena are explicable through variations of tactile sensation. In that, he agrees with Lewes (1878) and

James (1880). Bonnier maintains that "we have a sense of the attitude of any segment of our bodies by the sum of the tactile sensations in inter-muscular planes, in tendons, in articulations, and in the skin overlying them, and also perhaps in the muscle itself" (1393). The rejection of a specifically muscular sense leads instead to a consciousness of the effects at points of our body segments. Bonnier extends this argument, and considers special senses as variations of tactile structures, "an internal sense of touch," for example, the sense of orientation in invertebrates being derived from otoliths and their precursors, the "lateral line" in fish, a basic orientation system premised on the sense of contact with fluid (discussed further in terms of vestibular orientation and the semicircular canals of the inner ear in chapter 3).

By 1869, enough of a body of research existed that when Henry Charlton Bastian published his essay "On the Muscular Sense, and on the Physiology of Thinking" across four consecutive issues of the *British Medical Journal* that year, the seeds of an alternative to this term were being sown. Bastian, a friend of the great scientist and sociologist Herbert Spencer, was aware of the work of Fritsch, Hitzig, and his compatriot Ferrier. Bastian declared that "there is evidence to show that the brain is assisted in the execution of voluntary movements by guiding impressions of some kind," but that these impressions were distinct from the "ordinary cutaneous" sensations or of "deep sensibility" (1869, 463). But he rejected the Swiss physician Sigismond Jaccoud's position that "we are conscious of the impressions of the 'muscular sense,'" as our experience of coordinating any motor activity is unconscious, so any "unfelt" sensations are not "revealed in consciousness," and any inferred "muscular sense" must involve "unconscious sensory impressions" passing up to the cerebrum or cerebellum (Bastian 1869, 463; see also Wade 1972, 303). After 1870 Bastian "no longer had any doubt that the muscular sense ended in the motor area," according to Jeannerod, although he doubted the terminology of the "motor" aspect of the cortex, considering it more as a "*kinesthetic* center" (Jeannerod 1985, 62, emphasis in original). Bastian additionally devoted the entire appendix of his book *The Brain as an Organ of Mind* (1880) to "views concerning the existence and nature of a muscular sense" (1880, 691ff) and, having wrestled with the pros and cons of the term "muscular sense" within it, Bastian suggests an alternative term, "kinæsthesis." Of course, kinæsthesis (and its cognate "kinaesthesia") is a compound noun that adds movement *(kinein)* to the Greek root *aisthêsis,* together approximating a sense or feeling of movement. Bastian further establishes his new terminology in a later monograph-length paper published in the journal *Brain,* "The 'Muscular Sense'; Its Nature and Cortical Localisation" (Bastian 1887). Bastian's contributions on the "muscular sense" in these publications aid us in navigating decades' worth of physiological literature on the topic. But it is his contemporary Charles Scott Sherrington who provides a direct route both conceptually and physiologically between the sensory nerves of the muscles and the region within the cerebral cortex known as the motor cortex, putatively responsible for the planning, execution, and control of voluntary muscular movements.

SHERRINGTON AND "PROPRIO-CEPTION"

In the process of writing his essay on "The Muscular Sense" for Edward Schäfer's *Textbook of Physiology* (1900b), Charles Sherrington had prepared a number of microscope slides of nerve and muscle tissues, stained and ready for observation. Sherrington's materials, laboratory apparatus, and documentary archives are widely geographically dispersed. Despite one collection surviving at the University of Oxford since 1936, in 2010 a wooden box with his name attached containing twenty-one drawers of these slides was rediscovered. The rediscovery prompted an article in one of the *Nature* journals (Molnár and Brown 2010) to reassess his monumental legacy, which includes his early pioneering research in the spinal cord and motor control of dogs, primates, and then humans. Sherrington was awarded the Nobel Prize for Physiology or Medicine in 1932, jointly with Edgar Adrian. In his contribution to a coauthored seventh edition of a different physiology textbook he coined the term "synapse," from the Greek *synapsis,* to clasp together (Foster and Sherrington 1897, 929), and also coined "proprio-ception" in his major monograph *The Integrative Action of the Nervous System* (1906), elaborating on that concept further in an article for *Brain* the following year (Sherrington 1907). Some selected highlights of his research not only indicate the breadth of his scientific contributions, but also serve as an indicative conceptual pathway from a localized "muscular sense" within the body of previous sections to the mapping of the motor cortex, first in primates and then in humans. He identified the mechanism of inhibition (reciprocal innervation), whereby posture and coordinated movement in the animal as a whole occurs through a set of opposing reflexes, and wrote on the phenomenon of "decerebrate rigidity," whereby animals such as rabbits, dogs, cats, and monkeys maintained stiffness in their musculoskeletal frame upon removal of entire cerebral hemispheres. Sherrington assumed this was "due to spasms of the extensor muscles as a result of afferent stimulation from the sensory fibres in the muscle itself" (Molnár and Brown 2010, 432), conferring sensory as well as motor pathways within muscular tissue. Based on the rediscovery of the prepared microscope slides, specialized receptors (proprio-ceptors) were thought by Sherrington to be positioned throughout the muscular tissues and accessory organs (tendons, joints, blood vessels) in order to provide feedback on the organism's posture, position, and movement. While this posture and muscular control were coordinated from the brain via motor nerves, Sherrington was the first to identify afferent feedback, notes Burke in his centenary appreciation of Sherrington's book (2007, 890)—that is, sensory nerve impulses transmitted from peripheral muscle tissue back to the brain. In lecture 3, Sherrington considered how these afferents were "proprioceptive" because they are caused by the organism's own movements, in contrast with "exteroceptive" afferents that convey more distal information from the environment. The story of the discovery of proprioception and the ramifications for neurophysiology are difficult to summarize, given Sherrington's storied career and his many discoveries. This section can only offer selected waypoints in line with the narrative of sensation and interiority.

A few words should be devoted to explaining the centrality of the reflexes for the overall organization of the organism during the period Sherrington was maturating the ideas that would become *The Integrative Action of the Nervous System*. The story of the reflex itself encompasses figures including Hermann Lotze, Herbert Spencer, John Hughlings Jackson, and perhaps most famously, Sherrington's near contemporary Ivan Pavlov, who also won the Nobel Prize in Physiology or Medicine in 1904, specifically for his work on reflexes. For Spencer, one of Sherrington's influences, the reflex was an elementary or fundamental unit, regarded as "the atom of the psyche" (in Jeannerod 1985, 44). This was comparable to Sherrington's position that the "simple reflex" was the basic unit of neural function. But contra Spencer and Pavlov, Sherrington did not regard the nervous system as some adaptation of the organism to its external environment. Instead, the force for organizing and coordinating behavior must come from within the organism, so that the nervous system imposed its order on the environment, not vice versa. One thing Pavlov shared with Sherrington is the centrality of the cerebral cortex in the top-level organization and coordination of reflexes. Decortication (the removal of the cerebral cortex, as previously demonstrated by Flourens and others) he found would limit the acquisition of conditioned reflexes, and would suppress those that had been acquired before the procedure. As Jeannerod summarizes, prior to Pavlov's research, the identification of higher nervous centers with "superior" behaviors was orthodox, but after Pavlov the identification of higher nervous centers with the cerebral cortex more specifically was cemented, leading to the aphorism: "No conditioned reflexes without cortex" (43). For Sherrington, unlike Spencer, the reflex arc was the most basic element of the nervous system insofar as its function could be integrated and coordinated. Nevertheless, even in the introduction to his book, Sherrington acknowledges that the "simple reflex" is not the conceptually neat "atom of the psyche" that Spencer envisaged, and "is probably a purely abstract conception, because all parts of the nervous system are connected together, and no part is probably ever capable of reaction without affecting and being affected by other parts" (1906, 8).

The identification of such simple reflexes, just as "his mapping of sensory and motor pathways," explains Judith Swazey (1969, 131) in her book on Sherrington and reflexes, were therefore significant preliminary steps for understanding more complex reflex interactions. Simple reflexes are adjusted into an ordered "reflex pattern," and there must be orderly changes from one reflex pattern to another, a "co-ordination of reflexes successively proceeding" (Sherrington 1906, 8) from one to the other within the organism, and ultimately between the organism and its immediate environment. He extended the work of Kühne on the muscle spindle, identifying sensory fibers within the muscle tissue that provide feedback directly to the spinal cord for tonus (muscle position) and posture, proving definitively that muscle is in fact a sensory organ. As Molnár and Brown put it: "The discovery of sensory fibres in muscle spindles provided Sherrington with evidence that when a muscle is stretched, it sends proprioceptive information on the state of the muscle length to the spinal cord, which allows the animal to regulate its motor movements" (2010, 433). By identifying sensory nerves within muscle tissue, and tracing them from the spinal cord to the muscle spindles, he was able to define both the sensory and motor aspects of reflexes that originated in the spinal cord, and assess through experimentation which spinal nerves were affected when parts of the motor cortex were removed. Indeed, this experimental work led to the first maps of cortical function in primates (430).

In fact, Sherrington made his name in the 1880s by following up experiments on monkeys by Friedrich Goltz and David Ferrier on the localization of motor function in the cortex. A motor cortex had been conjectured by Fritsch and Hitzig in 1870, but further research was necessary to confirm whether motor functions were localized to particular cortical regions. At a meeting of the Physiological Section of the International Medical Congress of 1881, Ferrier and Gerald Yeo showed that a monkey of unnamed species with major lesions, in fact the removal of a section of the brain believed to be the left hemisphere of the motor cortex, demonstrated corresponding paralysis in body areas on its right-hand side. Incidentally, Ferrier noticed that touch remained largely separate from movement in the animal: "As to its tactile sensibility there is not the slightest sign of impairment" (Klein, Langley, and Schäfer 1883, 237; see also Finger 2005, 155–59, for a dramatic account). At that same meeting, Friedrich Goltz's experiment with a dog proved the opposite, that lesions in the same area revealed little paralysis or "sensorimotor deficit," as Molnár and Brown put it (2010, 429). At stake was nothing less than the experimental verification of the localization of motor function, with Goltz arguing for no localization, and Ferrier and Yeo arguing vociferously for it. The ensuing discussion, recorded in the *Journal of Physiology* by Klein, Langley, and Schäfer (1883), includes Yeo's skepticism that Goltz had removed enough of the dog's motor cortex, for the dog seemed so little impaired apart from some quirks in its behavior. It was then decided that Members of the Physiological Section of the Congress would be assigned to euthanize the animals and remove their brains to settle this, and the next day the meeting reconvened with seventy-five attendees. Of the preserved brains, observations were made concerning the relative completeness of motor cortex removal in the monkey brain whereas, as Ferrier and Yeo suspected, the lesions made by Goltz on his dog's brain were less precise, and "found to have more sensory and motor cortex spared than the professor had thought," according to Stanley Finger (2005, 158). Of course, history has remained on the side of the deputized committee in finding that motor functions were localized within a region of the cortex that spans across both hemispheres.[2]

A letter to the physician Theodore Dyke Acland of 1900 offers an intriguing snapshot of Sherrington's recent progress on mapping sensory pathways through the spinal cord. Presumably incorporating Acland's own phrase into his reply, Sherrington wrote: "The paths of sensory conductivities in the cord are 'in a fearful mess.'" Although the mapping was clearly still in progress, Sherrington described how nerves for touch, pain, and temperature shared common pathways in a section of the lateral column of the spinal cord, but are "separate and not commingled with the tactual" nerves (Sherrington 1900a). At that time, the cerebral terminus for those pathways was uncertain, possibly the "Rolandic region." The pathway for the "muscular sense" was more straightforward, apparently without the lateral crossing through the spine that other nerves exhibited, entering into columns of grey matter at the root of the posterior dorsal horn and ascending into the "superior worm of the cerebellum," an appropriately worm-shaped area of the cerebellum near the brainstem associated with bodily posture and locomotion, now termed the cerebellar vermis. Much of this was speculative, however: "Probably these paths are not concerned with 'muscular sensations' or 'muscular perceptions' really but by unconscious processes help regulate & help to coordinate movements—especially such habitual reactions as walking, posture, &c" (1900a).

In light of all this, Sherrington's entry "The Muscular Sense" for Schäfer's *Textbook of Physiology* (1900b), for which the microscope slides were prepared, allowed him to appraise the scientific literature on the controversial modality in the light of his own physiological experimentation. Like Bastian does across his several publications on this topic, Sherrington guides his reader through the somewhat technical and contradictory evidence. Right away Sherrington admits that a concise definition is elusive, as prior scientific investigators failed to employ the term consistently. Accordingly, he starts to assert his position rather broadly: "It may perhaps best be taken to include all reactions on sense arising in motor organs and their accessories" (1900b, 1002). Over the course of the entry, after considering the contributions of fellow investigators such as Bell and Bastian, he arrives at divergent conclusions. A few brief examples suffice. The role of the skin in informing the muscular sense is negligible, as "cutaneous anaesthesia inflicts little injury on the sense of movement or posture" (1011). Kirchner's argument, above, about the sense arising from the alterations and tensions of joints, muscles, and tendons is somewhat negated by Sherrington's observation that passive movement—that is, having one's limbs moved by someone else—produces no such sense perceptions. In the perception of muscular resistance, Sherrington allows that touch is "contributory," but the basis for judging different weights, for example, derives from "kinaesthetic" sensations that accompany "muscular action" (1021), at this stage following Bastian's terminology, although he will soon propose his own variant. More significant is the positive contribution Sherrington makes to the classification of the four types of percepts of the muscular sense—namely, posture, passive movement, active movement, and resistance to movement. Each type of bodily movement above a certain liminal level produces "changes in consciousness," he argues, that do not derive from single receptor types: "All these kinds of movements are perceptible by the mind, in virtue of the space perceptions with which it is furnished, by the combined labours of the senses of sight, touch, and the organs of the labyrinthine and muscular senses" (1013). The manifold nature of this muscular sense modality leads him to an observation that will be significant for his career two decades later, as I explore in chapter 5: "To the organs of muscular sense is largely traceable local 'feeling of fatigue'" (1021).

Although Sherrington's 1900 essay includes a section on "Perception of Posture," there is no mention yet of "proprioception," the sense of one's own bodily position, until his 1906 monograph *The Integrative Action of the Nervous System.* The term was subsequently elaborated upon in his article for *Brain,* "On the Proprio-ceptive System, Especially in Its Reflex Aspect" (Sherrington 1907). Between 1892 and 1894, Sherrington had conducted laboratory research on muscular reflexes, and these experimental observations formed the basis of his prestigious Silliman Memorial Lectures at Yale in 1904. The monograph was based on those lectures, and published two years later. The book, dedicated to Ferrier, was significant enough to be reprinted five times and, according to Swazey, even its last 1947 edition was sponsored by the British Physiological Society and deemed worthy of review in journals (1969, 127–28). Sherrington had built upon the work of Marshall Hall and Johannes Müller to more systematically develop the idea of the "reflex arc," a way that sensory stimulus and motor response remains localized between a part of the body and the spinal cord without having to relay centrally to the cerebellum. By not being transported the full extent of the nerve trunk, up through the spinal cord and into the brain regions, the speed of transmission inward to the spinal cord through sensory nerves (afferent pathways), and outward through motor nerves (efferent pathways) to effect muscular activity, is far faster. Sherrington had observed behaviors in the laboratory that suggested this arc was present even within primitive organisms. Yet what intrigued him was how the complexity of the many potential reflex pathways, what he sometimes calls "segments" or even "metameres" in the case of organisms like the earthworm (1906, 314), somehow effected an overall "functional unity," an oft-repeated phrase in the book (e.g., 1906, 77–79, 314), within the organism as a whole. Although some reflexes were relatively simple in themselves, others were more complex, and so could function as a coordinated unity through what he called the "integrative action" of the nervous system. From "simple reflexes" such as the knee jerk, to more complex pathways such as the scratch reflex of a dog's hind leg, with its coordinated sequences of nervous and muscular actions, the dog as a whole organism integrates the sum of such local reflexes and local actions into a larger coordinated whole. Furthermore, if the involuntary scratch reflex is based on local tactile stimulation, such as the alighting of a flea on the dog's fur, other senses—especially vision—are involved in coordinating the reflexes and their inhibitions into a series of voluntary movements, such as seeking and then moving to shelter when cold. Thus, for Sherrington even simple reflexes were significant and worthy of detailed physiological investigation for the way they became integrated:

> But the same principle extended to the reactions of the great arcs arising in the projicient receptor organs of the head, *e.g.* the eye, that deal with wide tracts of musculature as a whole, involves further-reaching coordination. The singleness of action from moment to moment thus assured is a keystone in the construction of the individual whose unity it is the specific office of the nervous system to perfect. (176)

These systems of reflexes are developed and then honed as patterns of behavior by the organism, and in the seventh lecture (235ff), Sherrington explains how reflexes have a "purposiveness" in terms of natural selection, having an adaptive or survival value for that organism. The study of reflexes as adapted reactions falls into one of two oppositional categories for Sherrington. First, as we have seen, flowing from simple reflexes and reflex arcs to the behavioral complexity of a whole organism. For an intact (i.e., neurologically complete, nondissected) animal, the complex interactions of simple reflexes throughout the nervous system "are ever combined into great unitary harmonies, actions which in their sequence one upon another constitute in their continuity what may be termed the 'behaviour' . . . of the individual as a whole," he says (237). In some invertebrate species of crabs or starfish, he observes, a continued disturbance by a predator might lead to a muscular reflex so pronounced that a limb is physically shed. Such behavior only underscores the extent to which reflexes are geared toward a larger integrity of the organism: "Such reactions exhibit well how absolutely the nervous system is adapted to minister to the requirements of the organism as an integrated whole, and the position of that system as a keystone in the upbuilding of the solidarity of the individual" (239). Second, and equally significant for Sherrington's own work, we might consider the "purposes" of various reflex types that have evolved within a vertebrate animal with intact spine by abstracting and breaking down the complex behavioral unity of an organism into its component reflexes: "In the analysis of the animal's life as *a machine in action* there can be split off from its total behavior fractional pieces which may be treated conveniently, though artificially, apart, and among these are the reflexes we have been attempting to decipher" (237, emphasis added). In other words, for the purposes of scientific observation, an organism's behaviors may be abstracted down into functional units of analysis as if they were a complex machine.

From a physiological perspective, then, reflex arcs serve as a convenient functional unit of behavioral analysis, separable from any localized activity in the motor cortex or cerebellum. With reflex arcs performing a range of behavioral functions, the prospect arises that some autonomy is wrested away from brain-centered conscious activity. But the densely composite nature of reflexes and arcs for an organism cannot tell the whole story. There must be a way to coordinate and trigger inhibitory mechanisms to suppress those reflexes that are unwanted or untimely, for a start. Furthermore, in order to accomplish tasks, there must be a coordination of posture and musculature as a result of sensory stimuli from an organism's external environment (what Sherrington calls "extero-ception"), such as audition, olfaction, and vision, via the "organs of the head," as he had put it. From this perspective, an organism's receptor organs initiate the reflex reactions in the nerve centers of each segment or metamere, and fall into two categories with different associated fields: the "deep" or proprioceptive field, and the surface field, comprising exteroceptive and interoceptive regions. An organism's capacity for exteroception, the arrangement of receptors for the perception of external stimuli such as touch and pain, leads to a far richer "extero-receptive field" than that of "intero-ception" (316ff), Sherrington explains, which has a dearth of receptors and is mostly limited to the processes of digestion. Unlike the intero-receptive field, the "extero-receptive surface" faces "outward on the general environment" of the organism, says Sherrington, and "contains specific receptors adapted to mechanical contact, cold and warmth, light, sound, and agencies inflicting injury *(noxa)*" (318). But whereas the intero-ceptive field is associated with the visceral or the alimentary, and the extero-receptive field is predominantly cutaneous (320), the third, proprioceptive field Sherrington associates with depth rather than surface is most pertinent to this chapter. This new modality based on neuro-muscular feedback relative to the organism has its own associated field, just like exteroception and interoception, of course.

Actually, Sherrington had postulated the necessity of "proprio-ception" only as a result of reflexes in complex organisms that had a "proprio-spinal tract," with proprio-spinal fibers reaching from the cerebellum down through the neck and spine and outward through the limbs, producing specifically "proprio-ceptive arcs" among the many other reflex arcs. But the centrality of the brain for the coordination of such proprioceptive arcs was never in doubt: "The cerebellum may indeed be described as the head-ganglion of the proprio-ceptive system," he asserts twice (iv, 347). For all reflex arcs, including proprioceptive ones, the brain has a paramount role in integrative action, as in the words of Swazey it "directs nervous integration in the normal animal," and in the first seven lectures, Sherrington "considers first the reactions of the motor cortex: its similarities to, differences from, and interactions with spinal centers in producing reflex movements" (1969, 151). In lecture 9, entitled "The Physiological Position and Dominance of the Brain," although it starts by summarizing his earlier points on the diffusion of the synaptic nervous system, and the role of functional segments and metameres, the centrality of the role of the cerebrum and cerebellum for integrative action is underlined. Given the relative poverty of the intero-ceptive field, he identifies the crucial role of "distance-receptors," the selection of receptors that provide a rich extero-receptive surface for an organism, which "seems to have peculiar importance for the construction and evolution of the nervous system," he says. *"The brain is always the part of the nervous system which is constructed upon and evolved upon the 'distance-receptor' organs"* (1906, 325, emphasis in original). What his discussion of proprioception and the proprioceptive system makes clear, based on his earlier neurosurgical interventions with the motor cortices of *Macacus rhesus,* is the way that postural and movement tasks between the motor cortex in the larger cerebrum are separated from the smaller, walnut-shaped component of the hindbrain known as the cerebellum (Latin for "little brain").

In addition to the previously described work of Ferrier, Yeo, and others on the motor cortex of dogs and monkeys, around the same time Luciani had performed ablations of the cerebellum, including its complete removal from a dog, publishing a series of papers in 1884 on its function that demonstrated disturbances in muscular coordination (*BMJ* 1884). These were not the earliest surgical demonstrations, as Legallois in 1812 had decerebrated a frog and noted how its movement was "disordered, without purpose" (in Jeannerod 1985, 46). Pierre Flourens had presented findings to the Académie de Sciences in 1822 on the results of a series of lesions, including removal of the semicircular canals in one pigeon, an entire cerebral hemisphere of another pigeon, and both cerebral hemispheres in another. The result of the most drastic surgical variant was "a loss of almost all sensation, and an inability to instigate purposeful behavior," according to Wickens. It remained motionless, yet its pecking reflexes remained intact when food was placed on its beak, but curiously retained the ability to fly when "when thrown high into the air" (2014, 152). But Sherrington's intention was to broaden the applicability of his findings to animal locomotion, and in Swazey's words, "explain the functional unity of motor behavior" (1969, 168). Although the idea of nervous integration preceded Sherrington, says Swazey, "the significance of Sherrington's concept lies in the fact that it provided the first comprehensive, experimentally documented explanation of *how* the nervous system, through the unit mechanism of reflex action, produces an 'integrated' or 'co-ordinated' motor organism" (168, emphasis in original).

At least in vertebrates, then, a picture was emerging of an entire assemblage that involved distance receptors oriented toward external environmental stimuli (exteroceptive field), and proprioceptive receptors for musculoskeletal position and posture, aided by the inner ear for spatial orientation (interoreceptive field). Bodily orientation remains relative to the position of the head segment, so the labyrinth of the inner ear is crucial, with the semicircular canals tracking spatial orientation and the otoliths movement and bodily inertia (this neuroanatomy and its history is developed further in chapter 3). These are coordinated chiefly through two brain locales, the motor cortex within the larger cerebrum, and the motor function of the cerebellum. With the cerebellum as the "head ganglion" of the proprio-ceptive system, as Sherrington claimed earlier (1906, 347), "the cerebrum [containing the sensory and motor cortices] is the head ganglion of the 'distance-receptors'" (349). At one point he speculated on a functional distinction between the motor cortices of the cerebrum and the cerebellum, such that the "cerebellum is the centre for continuous movements and the cerebrum for changing movements" (303). The labyrinth interacts with proprioceptors from the central trunk and peripheral limbs in order to maintain tonic reflexes—that is, stable bodily posture through skeletal muscular tensions—and ensures that the organism is always correctly oriented toward the outside world. It makes sense for Sherrington therefore to align the proprioceptors of the muscles, muscle spindles, and joints, offering postural feedback, with receptors within the labyrinth, which provide orientational feedback, all within the interoceptive surface. "The proprio-ceptors and the labyrinthine receptors," he says, have in common that "they both originate and maintain tonic reflexes in the skeletal muscles" (339). In order to coordinate this maintenance of musculoskeletal posture, the cerebellum does the work by integrating the "afferent contributions from the receptors of joints, muscles, ligaments, tendons, viscera, etc. . . . with those from the muscular organs of the head and with those of the labyrinth receptors themselves" (346–47). In a passage on the "total segmental attitude" of the organism, this vision, which includes the crucial element of proprioception, is effectively laid out:

The labyrinthine receptors and their arcs give the animal its definite attitude to the external world. The muscular receptors give to
the segment—e.g. hind limb—a definite attitude less in reference to
the external world than in reference to other segments, e.g. the rest
of the animal. Our own sensations from the *labyrinth* refer to some
extent . . . to this environment, that is, have some projected quality;
our *muscular* sensations refer to the body itself, e.g. contribute to perceptions of the relative flexions or extensions of our limbs. The arcs
of the proprio-ceptor of the leading segments control vast fields of
the skeletal musculature, and deal with it as a whole, while the arcs
of the proprio-ceptors of the other segments work with only limited
regions of the musculature. (343, emphasis in original)

The emerging cartography of brain regions and receptors that has
emerged from Sherrington throughout this section is fundamentally
based upon mechanistic assumptions of the interactions and integra-
tions of various reflex arcs and subsystems. Identifying such interactions allows him to venture explanations for historically observed
physiological puzzles, such as how vertebrate tonus is maintained, the
contribution of the vestibular system within the labyrinth of the inner
ear for an organism's posture and orientation, and "decerebrate rigidity," where removal of the entire cerebrum of living frogs led not to
muscular collapse, but surprisingly to the maintenance of muscular
tensions to maintain the overall postural "attitude" of the animal.

However, at one point in lecture 9, Sherrington steps back from the
details in order to offer a more panoramic view of the evolutionary development of nervous systems through the phylogenetic sequence. The
implication is that, for a range of anatomically sophisticated "higher"
animals, the sensory is invariably coupled with the motor, and so
"sensory-motor." He notes that "by its branching the motor neurone [*sic*]
obtains hold of many muscle-fibres" (309), and this diffusion of nerve
endings through muscle fibers feeds back to a ganglion within the cerebellum, what Lidell and Sherrington will later term the "motor unit"
(Burke 2007, 892). It is this same preconscious integration of movement
and muscle-fiber that allowed Dewey in his famous essay "The Reflex
Arc Concept in Psychology" of 1896 to speak of "sensory-motor coordination," uniting an initial sensory act or stimulus, say visually noticing

an object or person, with an associated movement to achieve an over-all action—for example, steering a car around a raccoon or waving to a friend. Dewey's article is a critical response to earlier formulations, where the sensorimotor circuit or system is interpreted from "precon-ceived and preformulated ideas of rigid distinctions between sensa-tions, thought, and acts," he says (Dewey 1896, 358). He thus departs quite radically from the prevalent idea in psychology, and we can see in Sherrington's work on integrative function that identifying a *struc-tural* unit can be straightforwardly converted into a unit of *function.* Identifying the appropriate mechanisms is very different from consid-ering the organism's own fused experience of perception and action:

> We begin not with a sensory stimulus but with a sensori-motor coor-dination . . . In a certain sense it is the movement which is primary, and the sensation which is secondary, the movement of body, head and eye muscles determining the quality of what is experienced. (358; see also Pfeiffer and Bongard 2007)

Yet, in a more philosophical mood, Sherrington's Rede Lecture deliv-ered at Cambridge entitled "The Brain and Its Mechanism," and pub-lished in 1933, speculates in rather general terms how the shaping of the animal's brain and body is a mutually beneficial collusion in order "to increase the animal's grip on the world about it" (1933, 9). There are analogies between animal brain function and motor function where the one stands in relation to the other like a key in a lock. It remains a deter-minedly mechanistic metaphor, one that downplays Dewey's formula-tion of reflex arc, or the possibilities that Goldstein ([1934] 2000) would indicate of plasticity, change, or function of the organism as a whole in the face of lesions or localized dysfunction. But when it comes to the localization of brain function, Sherrington goes on to argue, it does seem to matter where in the brain nerve processes take place. Whereas more commonly we speak of a distributed nervous system where brain function is not the only organizational agent, Sherrington speaks of the "nerve-net" that extends the centralized concept of the cerebrum out-wardly to encompass more of the animal's perception and grip on the world, and where even reflexes report back to the centralized cerebrum as conscious events with at least an approximation of a one-to-one relation: "This, one field of the nerve-net is visual, another auditory. Again, the position of a touched skin-point acts on the mind. The 'local sign' of reflexes has a mental counterpart" (29). He goes on: "The history of the brain seems to suggest mind as at outset a sort of adjunct to the nerve-management of motor behavior," so, for example, in eating, the motor behavior involves swallowing and chewing, but as it passes to the interior of the body "all mental experience of it ceases" yet "it does still continue to excite nerves and muscles, but the mind is in touch with it no longer" (29–30). In his preface to the final 1947 edition of his first great monograph, Sherrington weighs in upon one of his great philosophical interests, the relationship between the mind and the body. His book *Man on His Nature* (1940) was a late career philosophical excursus on this topic, including a discussion on the proprioceptive experience of limbs as "phantoms" based upon his earlier neurological evidence (1940, 308–9; quoted earlier in the introduction).

Returning from the philosophical excursus, this section ends with a final point from Sherrington that helps set up the following one. In lecture 8 on the motor cortex, he writes of the many labors of previous generations of physiologists in attempting to map the nervous system, and cautions the reader that since Hitzig and Fritsch, Ferrier, Hughlings-Jackson, Bastian, and others, discoveries have proceeded according to assumptions of localization. But Sherrington warns us against overly simplistic interpretations of brain maps, as the brain, and particularly the motor cortex, may not yield so easily:

> The discovery of localization of function in parts of the cortex has given the knowledge which now supplies to the student charts of the functional topography of the brain much as maps of continents are supplied in a geographical atlas. The student looking over the political map of a continent may little realize the complexity of the populations and states so simply represented. We looking at the brain chart of the text-book may never forget the unspeakable complexity of the reactions thus rudely symbolized and spatially indicated. (Sherrington 1906, 270)

It is fair to say that this advice remained unheeded by the next cartographer of the motor cortex, Wilder Penfield.

The Motor Cortex: Penfield's Cartography and Cantlie's Creature

There is no doubt that the 1870s were a signal decade in establishing the viability of the project of identifying, and then through direct electrical stimulation, topographically mapping the motor cortex. As we saw, key experiments involving the direct stimulation of dogs by Hitzig and Fritsch in 1870 and then Ferrier's cortical lesions on monkeys in 1874 confirmed the existence of a motor cortex directly associated with bodily movement and its planning. Hitzig and Fritsch showed how certain areas of the cortex were "excitable" and, through the application of platinum electrodes with galvanic current directly onto the cortical surface, were able to elicit weak sensations, "establishing the first functional map of the cortex based upon objective findings" and from which "the idea of a cortical center for motor activity spread rapidly" (Jeannerod 1985, 55). Placing their electrodes on the anterior region of the cortex, the current produced a contraction in a muscle group on the opposite hemispheric side. Contemporaneously, the British neurologist John Hughlings Jackson was studying epileptic patients and similarly identified the existence of a motor cortex located on the surface of the cerebral hemispheres. Autopsies later confirmed degeneration in these cortical areas, and on the opposite side of the body's affected parts. Ferrier sought to further Jackson's observations and hypothesized that, after Hitzig and Fritsch's technique, direct electrical stimulation would produce convulsions in monkeys that could be observed systematically. Whereas galvanic current produced only brief contractions, he found that faradic current produced contractions of longer duration that, he believed, more closely resembled the "intentional" movements of the primate, such as grabbing an object in front of it (55). However, Ferrier's map of the motor cortex was incorrect and ended up extending far beyond its actual boundary, well into the posterior zone that would later be associated with the somatosensory cortex. In the wake of this, in a series of experiments involving electrical stimulation from 1901, Sherrington and Grünbaum urgently sought to establish the definitive topography of motor areas in a series of "higher" primates, including the chimpanzee, orangutan, and gorilla, culminating much later in a paper with Leyton (Leyton and Sherrington 1917).

Significantly, given what follows, Hughlings Jackson "conjectured that the motor cortex must be organized along somatotopic lines, so that the hands, face, tongue, and feet, which possess the greatest capacity for specialized movement, were correlated with larger areas of the motor cortex than other movable parts of the body" (Bennett and Hacker 2008, 230). Ferrier's work also encouraged the American neurosurgeon Harvey Cushing to experiment further by mapping the motor parts of the cortex between 1902 and 1912. And it was by building upon this work of Hitzig and Fritsch, Ferrier, Sherrington, and a surgical apprenticeship with Cushing in particular, that led American-born neurosurgeon Wilder Graves Penfield to set up the Montreal Neurological Institute (MNI), first to continue his work on epilepsy and then a larger mapping project, performing a series of neurosurgical operations from 1929 to create "maps" of bodily sensation and motor control by direct electrical stimulation of the adjacently situated somatosensory and motor cortices, respectively.[3] In the second chapter of their book, *The Cerebral Cortex of Man: A Clinical Study of Localization of Function* (1950), entitled "Sensorimotor Representation of the Body," Penfield and his collaborator Theodore Rasmussen briefly recap the progress of predecessors such as Hitzig, Ferrier, Jackson, and Luciani on the boundaries and limitations of a motor area, and whether it was "sensorimotor" rather than strictly "motor." Having developed a form of craniotomy that allowed access to the top surface of the cerebrum, Penfield concentrated this phase of his research on a landmark fissure thought to demarcate the motor cortex and the somatosensory cortex, laying the foundations for what is now called the primary and secondary somatosensory cortices (S1 and S2). From any time in the second half of the twentieth century to today, if you were to open any psychology textbook at the section on the brain, it is highly likely you would see a version of one of Cantlie's illustrations of Penfield's cortical maps. As Richard Griggs wrote four decades ago: "Look for *homunculi*—those strange drawings of distorted human anatomy in which the size of the body part is proportional to its area in the motor or somatosensory cortex and not its actual size" (1988, 105, emphasis in original).

The point of this section is not to rehearse the history of neuroscientific investigations into the motor cortex, however. True, Penfield does mention the "muscle sense" in passing in a brief historical note to a coauthored paper: "Nothnagel (1873), Hitzig (1874), Schiff (1875) and Hermann (1875) thought that the cortical centres were really sensory centres for 'muscle sense'" (Penfield and Boldrey 1937, 390). But the main focus of his research was, first, to pioneer surgical techniques for cortical mapping, and second, to call wider attention to his mapping project by collaborating and publishing with a series of visiting or trainee neurosurgeons, and by commissioning a medical artist to produce rather controversial illustrations of localized brain function. The resulting illustrations appeared in his published books and articles over nearly two decades, starting with his coauthored paper with Edwin Boldrey, "Somatic Motor and Sensory Representations in the Cerebral Cortex of Man" (1937), then a more comprehensive series of illustrations in his coauthored monograph with Rasmussen, *The Cerebral Cortex of Man* (1950). The monograph on epileptic seizure patterns by Penfield and Kristiansen, *Epileptic Seizure Patterns: A Study of the Localizing Value of Initial Phenomena in Focal Cortical Seizures* (1951), described "initial sensori-motor phenomena" in epileptic patients but included none of Cantlie's illustrations. The coauthored monograph with Herbert Jasper, *Epilepsy and the Functional Anatomy of the Human Brain* (1954), was the last to do so, with three sets of illustrated homunculi. The infamous "sensory homunculus" emerged and mutated through a series of sketches and collaborations by Penfield with the artist Hortense Pauline Cantlie (née Douglas) then, and the ongoing nature of the series offers glimpses of not only the process of mapping the cortices but also the limits of the forms of graphical representation of motor cortex function. In this section, the story of how such detailed neurosurgical work led to the "very strange-looking beings" (DeSalle 2018, 28) reproduced in many undergraduate psychology textbooks, the famous "sensory homunculus" and "motor homunculus," is explored. There are two aspects of this collaboration to be developed in this section. First, the rather unique style of art–science collaboration that occurs here, and the innovation of representing complex neurosurgical data in a simplified form. Collaborations between scientists and artists are not historically new, and other pairings occur in later chapters. But the reach and extent of this particular collaboration is notable given how widely reproduced and reprinted the illustrations are, and given how the role of the artist, Cantlie, has historically been downplayed. A paper by Griggs (1988), with the unusual title "Who Is Mrs. Cantlie and Why Are They Doing Those Terrible Things to Her Homunculi?," surveys a series of psychology textbooks to discern inaccuracies and errors in the reproduction of the images that Cantlie had originally made for the Penfield and Boldrey publication of 1937. However, the first part of the question that Griggs asks, as to the illustrator's identity, remains only sketchily answered. To this end, archival research at the MNI in June 2016 allowed me to examine an entire series of original ink drawings for Penfield's publications by Cantlie, and thereby see some results of this collaboration in detail. Second, there is the recent revisitation of Penfield's data that produced the illustration of the homunculus in the first place by Paula Di Noto and her coauthors from the University of Toronto (2013). Based on their findings, it became clear that Penfield's map of the body in the brain is almost exclusively male, and that data from female patients had been systematically excluded. Di Noto's team retrieve what they term a "*her*munculus" from the data (Di Noto 2013). Therefore, in addition to the issue of the potential erasure or distorted acknowledgement of Cantlie's graphic contribution to Penfield's work, the story of Penfield's somewhat cartoonish illustration has a troubled relationship with women's bodies and, we will find, female sensation.

THE BODY IN THE BRAIN: PENFIELD AND CANTLIE'S "GROTESQUE CREATURE"

Penfield and Cantlie's sensory and motor homunculi serve as one of the most startling instances of the concept of neural representation in the first half of the twentieth century, although Cantlie's contribution as an illustrator is rarely acknowledged, and the role of Penfield's first collaborator Boldrey was instrumental in its conception. A controversial and now discredited instantiation, like an exaggerated caricature-creature, it was birthed in 1936 when Boldrey, as a neurosurgical resident under Penfield, wrote his master's thesis "The Architectonic Subdivision of the Mammalian Cerebral Cortex, Including a Report of Electrical Stimulation of One Hundred and Five Human Cerebral Cortices" (Boldrey 1936). The origins of the schematic of the homunculus can be traced to Boldrey's data. According to Gandhoke et al., Boldrey's thesis "documented the first large-scale study of human in vivo cortical stimulation under local anesthesia during surgery for epilepsy and brain tumors and was the first such study to actively identify and separate motor and sensory functions, as well as other brain functions" (2019, 377). It included twenty-eight charts that contained a number of prominent stimulation points that helped formulate an overall sensory and motor sequence along the cerebral surface, the better to provide evidence of the localization of brain function. Since Boldrey was conducting the stimulation experiments in Penfield's surgical theater at the MNI, they codeveloped a methodology where the numbered points of cortical stimulation that marked different functional regions were then mapped directly onto a life-sized half-brain preprinted diagram in the theater. Drawing numbers on the diagram "was thorough yet not without occasional subtle approximations," as Gandhoke et al. (378) remark. These basic maps of a brain hemisphere with numbered points were used in Boldrey's 1936 thesis and also Penfield and Boldrey's 1937 paper. Arguably, however, the prototype of the homunculus was present in Boldrey's thesis where, as Gandhoke et al. (388) claim, "It has a form only of words describing an idea, but it is there, his 'abstract thing': a list of functions, both motor and sensory, in a sequential, typed, but spatial layout." Unlike the hemispheric point maps, having a list of functions typed out as a motor and sensory sequence helped conceptualize the data differently. It prompted Penfield to enlist the artist Hortense Douglas Cantlie (1901–79) to provide illustrations for his further experiments with Boldrey, expanding from 105 to 163 epileptic patients. In these operations, as before, points on both the motor and somatosensory cortex received electric current, and any resulting muscle twitches or local bodily sensations were systematically noted and represented. Verbal reports or nonverbal involuntary actions from stimulating small regions of the still-conscious patient's cortices were still pinpointed on the same hemispheric diagrams. But the 1937 paper also shows that small numbered labels ("tickets of reference," 1937, 426) were actually placed onto the cerebral surface (see Figure 2). In both Boldrey's thesis and Penfield and Boldrey's paper there is acknowledgement of the wide variability in response that the stimulation of a discrete cortical point could produce across different subjects. Any mapping between the point of cortical stimulation and the affected bodily area varied from patient to patient, and even from one operation to another for one patient in particular. Nonetheless, the illustrated homunculus that emerged from Cantlie's discussions with Penfield aimed to capture useful generalizations regarding the nature of the body's representation in the brain for the widest audience. One such generalization was that certain body parts, such as the thumb and tongue, always had a disproportionately large cortical representation. In the 1937 coauthored paper by Penfield and Boldrey where the homunculus first appears as an illustration, they themselves describe it as a "grotesque creature,"

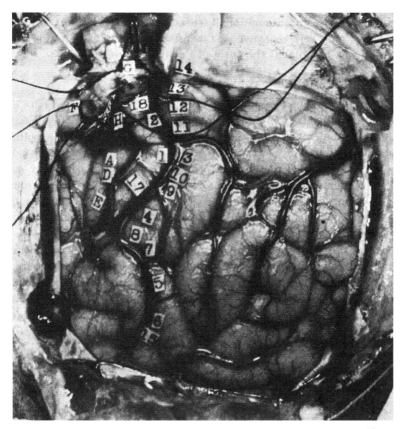

FIGURE 2. *The cerebral cortex with "tickets of reference" relating to functional regions, from Penfield and Boldrey (1937, 426). Reprinted with permission from Oxford University Press.*

and Penfield's contemporary Francis Walshe called it "a rather deceptive monstrosity" in a 1943 letter to Penfield (Walshe 1943). It is such inaccuracies, the "slightly ridiculous, cartoonlike—and by extension, unscientific" aspects as Guenther (2015, 156) puts it, that have made the figure not only controversial but also something of a historical curiosity within the scientific community (see, for example, Schott 1993; Catani 2017; Gandhoke et al. 2019). Its status as a familiar yet quasi-scientific caricature comes out in DeSalle's popular science book:

> Humans have a bizarre-looking sensory homunculus. . . . His hands look huge, his lips and tongue are supersized, and his penis is, well,

really gigantic. He has extremely large feet and a torso that looks ema-
ciated. This strange-looking being indicates that a large proportion of
the sensory strip in the human brain is dedicated to our digits, lips,
and genitalia. (DeSalle 2018, 29)

Since Griggs does not answer the first part of his question, "Mrs.
Cantlie" was in fact Hortense Douglas Cantlie, born in Yonkers, New
York, but raised and schooled in Montreal. She studied at the Montreal
Art Association in 1921 and in New York the following year. She studied at Johns Hopkins University under Max Broedel from 1925 to 1926,
where she obtained a certificate in Art as Applied to Medicine in 1926.
Between 1924 and 1935 she worked as a medical illustrator, principally
at the Montreal General Hospital. Her illustrations were used in medical
articles and books, but also the kinds of three-dimensional anatomical
wax models used for medical teaching in the tradition of the "moulage,"
popular from the nineteenth century. This includes a life-size (30.5 x
25 cm) rendering of a syphilitic lesion on the face of a twenty-seven-
year-old Chinese male, made circa 1924–35.[4] In their rather exhaustive
book on the history of the MNI, *The Wounded Brain Healed,* Feindel and
Leblanc explain that in 1932 Penfield received a postcard with Ludwig
Edinger's illustration of a brain with convolutions represented as ba-
bies nestling into each other, and was impressed with the design (2016,
258). He commissioned Cantlie to produce a version in stone for the
dedication plaque of the McDonnell Wing at the MNI in 1953 (see Figure 3), titled "Children of the Brain." These designs are rather fanciful,
Escher-like renderings of repeated patterns of bodily figures aligned
to show the shape of the brain. Penfield asked for slight modifications
from Edinger's original design so that the infants are posed according
to functional areas of the cortex. An infant cupping its hand on its ear
is placed in the area of the auditory cortex, for example. However, they
are not the kinds of representations of the body within the brain that
Cantlie's most famous illustrations, the sensory and motor homunculi
with Penfield, aimed for. That particular illustration has been reproduced in countless psychology and medicine textbooks and modelled
as a three-dimensional figure, too.

As we saw, the first schematic of a homunculus is possibly Boldrey's
1936 thesis, but the earliest illustration by Cantlie appears in the paper
by Penfield and Boldrey, "Somatic Motor and Sensory Representa-

FIGURE 3. *Brain Children, dedication plaque of the McConnell Wing at the Montreal Neurological Hospital (1953), designed by H. Cantlie, 1953. Photograph by author, 2016.*

tions in the Cerebral Cortex of Man" (1937, 432) where they explain its purpose:

> The homunculus gives a visual image of the size and sequence of cortical areas, for the size of the parts of this grotesque creature were determined not so much by the number of responses . . . but by the apparent perpendicular extent of representation of each part when these responses were multiple for the same part. (Penfield and Boldrey 1937, 431–32)

As can be seen (Figure 4), the thumb and lips are exaggerated in size, indicating that a disproportionately large area of the cortex will yield responses in those parts when electrically stimulated. This is technically untrue, and even Penfield avoided asserting this. Having become subject to criticism from his peers, the grotesquery continued in different forms, as the figure was refined in Penfield's later publications, including a coauthored book with Rasmussen in 1950 (Figure 5) and a coedited book with Jasper on epilepsy in 1954. For the 1950 version, Cantlie's illustration was now split into two, with a sensory (left) and motor (right) homunculus as a figure draped over a cross-section of the cerebral hemispheres, an attempt to more clearly associate cortical area

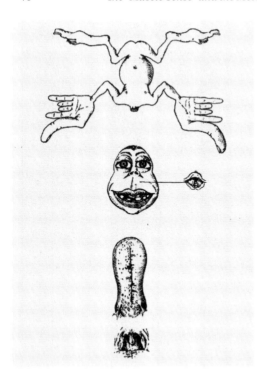

FIGURE 4. *Original 1937 sensory and motor homunculi of Penfield and Cantlie, reprinted in Penfield and Boldrey (1937, 432). Reprinted with permission from Oxford University Press.*

with sensory and motor function (Penfield and Rasmussen 1950, 44, 57). Feindel and Leblanc acknowledge Cantlie's contributions to both the original 1937 version and then the revised homunculus in Penfield and Rasmussen's 1950 monograph, honoring her married surname:

> Hortense Douglas-Cantlie illustrated the book with line drawings that are striking in their clarity. The whimsical full-frontal homunculus as originally drawn by Cantlie for the 1937 paper [by Penfield and Boldrey], upside down save for the face, reappears one last time, to be replaced by sensory and motor homunculi in profile, laid upon cross-sections of the hemispheres. There were other attempts at illustrating the homunculus, some more successful than others, but these images have reached iconic status and are indelibly imprinted in the mind of all neuroscientists. (Feindel and Leblanc 2016, 206)

In the legend for this revised sensory homunculus, Penfield and Rasmussen explain the underlying logic for this particular graphical form, which

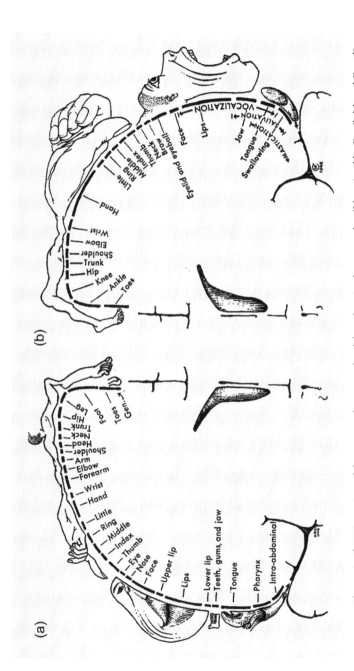

FIGURE 5. The revised 1950 version of the sensory homunculus (a) and motor homunculus (b) revised by Penfield and Cantlie and published in Penfield and Rasmussen's 1950 monograph on pages 44 and 57, respectively. Images from Cengage Learning Inc. Reprinted by permission. www.cengage.com/permissions.

explicitly entails topographical order as well as an indication of comparative quantity of response in the stimulated body parts: "The right side of the figurine is laid upon a cross section of the hemisphere, drawn somewhat in proportion to the extent of sensory cortex devoted to it. The length of the underlying block lines indicates more accurately the comparative extent of each representation" (Penfield and Rasmussen 1950, 44).

On paper, and in the many textbooks that feature it, the homunculi illustrations apparently depict a one-to-one correspondence between a location in the brain and a body part and either a movement (motor cortex homunculus) or a felt sensation in a specific part of the body (sensory homunculus). What it does not show, but which is apparent in the longer text and the inclusion of numerous data tables along with it, is that the grotesque creature can only emerge by virtue of careful cherry-picking of those data. So how did Penfield and his surgical collaborators end up with these representations? As with the more primitive cortical mapping on mammals of Hitzig and Fritsch from a century earlier, and the early experimentation on dogs with Goltz and monkeys by Ferrier described earlier and which Penfield and Boldrey mention briefly in a historical note to their paper, their homunculus is employed in support of assumptions of strict localization. The overemphasis on localization comes through in the way that the homunculus is depicted in both versions, with the 1950 version as if draped over the cortex, retaining some artistic license. The revised version of the homunculus was in fact more clearly mapped onto the object of Penfield's series of surgical investigations, what Penfield and Rasmussen called the "sensory and motor strip" that forms a "longitudinal fissure" (1950, 209) across the top of the cerebrum, and which he and his collaborators subjected to point-by-point direct electrical stimulation in open-brain surgery at the MNI from 1929 onward. In other words, through systematic direct brain stimulation of the surface of the cerebrum there emerged a functional map of those cortical areas concerned with sensation, and their bodily location. Descriptions of somatic sensory areas within the cortex appeared shortly after 1929, with a paper authored solely by Penfield (1930) in the journal *Brain*. But Penfield's 1936 commission for Cantlie for the coauthored paper with Boldrey published a year later introduced the homunculus to depict the whole sensory and motor sequence across the fissure in a graphically simple way. In both its early and later incarnations, it clearly shows the progression of the direct cortical representation of limbs, the trunk, the face, then passing inside through the mouth and the alimentary tract, where the sequence tails off. Unlike Sherrington's division of interoceptive and exteroceptive surfaces, both the interior and exterior sensations of the body were depicted along a single, continuous surface that happened to align along a fissure across both cerebral hemispheres. Therefore, when Penfield and Rasmussen wrote of "somatic sensation," this included "localizing sensation over the whole surface of the body, the mouth, and the tongue, and sense of position of those parts of the body that are subject to voluntary control" (209). Initially, like Ferrier, Penfield used galvanic (direct) current, but then developed a technique by supplementing it with faradic (alternating) current: "The first was used to map out the localization of sensory and motor function on the brain, while the second was used to elicit light epileptic seizures that would help confirm the focus of the offending brain tissue," explains Guenther (2016, 292). From 1935 to 1945, he incorporated a thyratron stimulator by which means he could more precisely control the frequency and amplitude of the current, and from 1945 a Rahm stimulator modified by Jasper (Penfield and Rasmussen 1950, 6).[5] As with the earlier surgeries with Boldrey, stimulation of the brain would be recorded with a number on a map, with numbers written on sterilized paper and physically placed on the cortical surface and photographed. The elicited motoric or verbal responses were noted and filed in a "Stimulation Report." For the surgeries with Rasmussen, they explained that "a secretary sits in the viewing stand, separated completely from the operating room by glass, and the operator dictates to her the results of the positive stimulations" (7). Having developed and refined this technique initially from 1929, Penfield could see the potential for a more general project in the vein of Hughlings Jackson's somatotopic map of the cortex. Thus in the paper with Boldrey of 1937, this was put in terms of producing a "standard human chart [of the brain] on which are marked cytoarchitectural fields" (Penfield and Boldrey 1937, 425)—that is, the spatial arrangement of different cell types with specializations of function in different brain regions.

As Penfield and Rasmussen themselves discuss, the series of craniotomies were difficult and time consuming to set up, but produced a large quantity of data regarding sensory and motor responses to electrical stimulation of the cortex. Since patients were only provided local

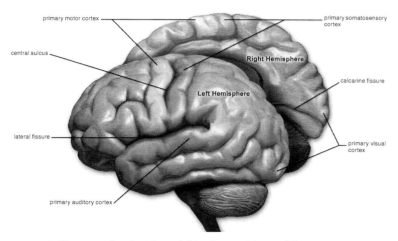

FIGURE 6. *Diagram showing the neighboring positions of the primary somatosensory cortex and primary motor cortex, separated by the central sulcus (formerly Rolandic fissure). Blausen.com (2014) "Medical Gallery of Blausen Medical 2014,"* WikiJournal of Medicine *1, no. 2. https://doi.org /10.15347/wjm/2014.010. ISSN 2002–4436. Creative Commons 3.0 license.*

anesthetic, this meant that the conscious patient's subjective response to cortical stimulation could be charted, its nature was far more precise and detailed than the cataloging of outward behavior of frogs, dogs, or monkeys in earlier cortical lesion experiments, and the subjective experience of the patient themselves was an inescapable and somewhat inconsistent factor. Furthermore, to impose order upon the quantity of data was inevitably to employ an abstracted form of mapping according to brain region and motor or sensory function, further compromised by Penfield's need to find representation in graphical form, the homunculus. Given the various means of indicating cortical localization, through maps, tickets, a hemispheric model in the theater, and glass separating the secretary writing the stimulation report, inevitably there would be inaccuracies. What Penfield and his contemporaries termed the Rolandic fissure, now known as the central sulcus, is a distinctive landmark on the surface of the cerebral cortex that demarcates the primary motor cortex of the frontal lobe from the primary somatosensory region in the neighboring parietal lobe (see Figure 6). At the beginning of this mapping project, Penfield (e.g., Penfield and Boldrey 1937) assumed that this fissure neatly separated the cortex into sensory and motor areas, and if cortical functions were as localized as the illustration assumes, there would indeed be functionally distinct motor and sensory homunculi. But actual data muddied the picture somewhat. For example, movements could still be elicited from stimulation of the somatosensory cortex (posterior to the Rolandic fissure and motor cortex), and vice versa, stimulation to the motor area anterior to the Rolandic fissure sometimes generated bodily sensations. In the face of these data, Penfield modified his position:

> The study of the cerebral cortex of man indicates that upon the precentral gyrus is to be found the primary representation of movement, but that there is a subordinate motor representation in the postcentral gyrus. Conversely, the primary somatic motor representation is postcentral but there is a corresponding representation of sensation in the precentral gyrus. (Penfield and Rasmussen 1950, 13)

Stimulus points were most effective around 1 cm each side of the Rolandic fissure to elicit either motor responses on the anterior side or sensory responses on the posterior. But now he acknowledged that for motor responses "approximately 80 per cent of these points were precentral in location [i.e., associated with the motor cortex] and 20 per cent postcentral [sensory cortex]" (46), and for sensory responses the division was 75 percent versus 25 percent (21). Other surgical errors and discrepancies further muddied the data, as Guenther (2016) shows. Yet, curiously, despite the growing evidence that brain functions were not so cleanly localized, he continued to have faith in the communicative power of his co-created sensory and motor homunculi. That is, until his interest developed in eliciting speech through direct stimulation, and then returning to his earlier project on epilepsy.

Sir Francis Walshe, a neurologist of the National Hospital in London, was one of the harshest critics not only of the homunculus but also of Penfield's work in general. The themes of Walshe's criticisms were, first, that electrical stimulation experiments cannot reveal the true functional organization of the cortex, and second, that the strict localizationist theory is a mistake that could be avoided if more attention were paid to the writings of the British neurologist John Hughlings Jackson who, like Penfield, had conducted extensive work on epilepsy but was critical of strict localization. For example, Walshe attacked not only Penfield's conceptual assumptions but also his methodology for their verification:

> The method of punctate electrical stimulation inevitably led to the evocation of small muscular movements, what Sarah Tower was later to call "discrete movements," and thus to a concept of what Jackson called "abrupt geographical localizations" and to the "cortical mosaic" of Fulton. Jackson maintained that no hypothesis of an integrative function for the cortex could be built on such foundations. (Walshe 1961, 120)

Regarding the homunculus, Walshe was equally critical. At one point in their correspondence, the unusual proportions of the homunculus were questioned. Penfield wrote back, downplaying their significance, emphasizing instead its value in calling "attention to certain facts, such as the reversal of order of representation in the face and neck, as compared with the rest of the body" (quoted in Guenther 2015, 157). Yet in the paper with Boldrey in 1937, one of their principal claims was how the figure illustrated the "size and sequence of the cortical areas" (1937, 431–32)—that is, as Guenther explains, accuracy in "both the order and proportion" (Guenther 2015, 157). It was in the face of the continued criticism, principally from Walshe, that Penfield relented and commissioned Cantlie to produce the second version of the figure, admitting in his 1950 monograph to "minor inaccuracies now apparent in this figurine" (Penfield and Rasmussen 1950, 26). Effectively, says Guenther, "Penfield was thus simultaneously deeply invested in the homunculus and seemingly at a loss for a workable justification of it" (2015, 157). In a letter from 1946 that responds to some of Walshe's criticisms, Penfield gives a candid assessment of his homunculus:

> It was ... one of a number of illustrations which we used to try to illustrate the truth. Of course, there is nothing like the homunculus as far as cortical representation is concerned, but it seems to be the only sort of thing that people in general understand. I would gladly kill the damn thing if I could, but that is never possible. (Letter to Francis Walshe, August 20, 1946)[6]

Walshe remained unimpressed, and even with the addition of time remained vehement in his criticism not only of its inaccuracy but also of its unnecessary grotesquery: "Nor are the moderns content with maps, for *homunculi* and *simiusculi* have now made their horrid appearance, lineal descendants of Lewis Carroll's Jabberwock, purporting to depict the fair face of nature, but in fact achieving something quite unnatural" (Walshe 1958, 232). Walshe elsewhere argued that the error of overreliance on stimulation experiments and the neglect of careful clinical observation is compounded by the representation of the results of such experiments in graphical form, including the standard genre of brain cartography of the sort that Sherrington had employed in his diagram of the motor cortex of the chimpanzee (Walshe 1953, 26; see also Sherrington 1906, 274). Walshe's criticisms were not alone. At the time of Penfield and Boldrey's initial paper in 1937, it was already being argued that stimulation experiments do not support the thesis of strict localization, and that alternative styles of graphical representation could better convey the functional organization of the cortex, with its anomalies and exceptions to strict localization. Consider Otfried Foerster's 1936 paper on Hughlings Jackson and the motor cortex, for example. Before arriving at Montreal in 1929, Penfield had studied under Foerster at Breslau in 1928, but his former teacher's words suggest a very different imagination of how brain data might be represented:

> The inconstancy of the effects and the variation of the responses of one and the same spot to repeated stimulations can be demonstrated in almost every focus of the precentral convolution. Each focus contains not only elements of the part of the body represented preponderatingly in that focus, but also elements of other parts of the body. The foci of the different parts of the body are not like the stones of the mosaic to which they were compared, but they overlap to a more or less considerable degree. The anterior central convolution does not resemble a painting of cubistic style, it reveals rather the intimate mixture of soft colours and smooth forms of a Raphaelitic Madonna. (Foerster 1936, 142–43)

The question of how pioneering Penfield and his collaborators actually were is open to question. Fedor Krause referred to Sherrington's earlier pioneering technique of faradic stimulation with monkey brains in his multivolume monograph *Chirurgie des Gehirns und Rückenmarks nach eigenen Erfahrungen* of 1908–11, and in fact conducted his neurostimulation on human epileptic patients not through craniotomy like Penfield but through a different surgical technique known as trephining, where a circular flap of skull is opened up and can be closed. Apart from this, Krause's methodology was very similar to Penfield's. For example, the process of pinpointing on a hemispheric surface the results of electrical stimulation was actually performed by Krause two decades prior to Penfield. In the second volume of Krause's monograph, translated in 1912 as *Surgery of the Brain and Spinal Cord Based on Personal Experience,* he clearly states his method: "Observations of muscular contractions of face and extremities are now dictated to the recorder, while the located centres are noted on a sketch of the cerebral surface by the artist" (Krause 1912, 293). These sketches, some of which are reproduced in his monograph, appear very similar in style to Penfield's hemispheric point maps discussed above. Further, in his repeated cortical stimulation, Krause even gets the upside-down nature of the motor-sensory strip along the cortical surface that Penfield had also found and had illustrated in the first 1937 version of the homunculus: "The arrangement on the cortical surface is such, that the centers for the lower extremity occupy the upper-most portion of the convolution near the sinus longitudinalis," says Krause (294). It is therefore possible that Krause's neurosurgical explorations invoked a proto-homunculus that predates even Boldrey's. Although Penfield's research with Boldrey produced landmark results, it is puzzling that Penfield there and elsewhere omits references to prior mapping research, including both Horsley and Krause's electric stimulation of the motor strip, as Gandhoke et al. (2019) point out.

The homunculus, it could be argued, was a necessary but insufficient step from the identification of mechanisms involved in the theories of the reflex arc and integration from Sherrington and others. This has been a shift from, first, an imagination of the sensory-motor body with sensory nerves located in muscle spindles, joints, and tendons; and second, a reflex model with the terminus of sensory and motor nerves in the spinal cord, coordinated in the motor cortex and the cerebellum; to, third, a novel way of continuing the work of sensory and motor mapping through direct stimulation of the sensory and motor cortices, the generation of a cortical schema and therefore an assumption of an interface between mind and body that is mechanical and manipulable. Surprisingly, the homunculus has made a recent return of sorts in the twenty-first century.

THE *HER*MUNCULUS: THE FEMALE BODY IN THE BRAIN

Errors and ambiguities in Penfield's surgical methods and data collection were mentioned above, but there are other concerns and omissions regarding the formation of Penfield and Cantlie's homunculi. Significantly, both versions of Penfield and Cantlie's homunculus illustration have male genitalia, and at no point does Penfield or his collaborators offer the view that the representations of the brain in the body are anything other than static, male, and normative. It turns out there are significant differences and discrepancies between the somatotopic maps of male and female subjects. There are also important variations of these somatotopic maps within male and female subjects due to factors including fluctuation of hormone levels, menstruation, and changes during the life course. A team of female researchers from Toronto spanning psychology, communications, and public health attempted to answer the question enshrined in the title of their article in *Cerebral Cortex:* "What Is Known about the Representation of the Female Body in the Brain?" (Di Noto et al. 2013). Their recommendation was to use more recent medical technology to measure evoked potentials in the somatosensory cortex with electroencephalography (EEG) combined with functional magnetic resonance imaging (fMRI) to produce a much more detailed somatotopic map of both male and female subjects, and therefore produce more accurate homunculi. They identify an urgency in redressing the comparative absence of data from female subjects and so produce a *her*munculus. There are two main aspects of their research that pertains to my narrative so far.

First, the paucity of female subjects, and therefore the omission of female anatomy, in Penfield's data over the course of his career at the MNI. Despite one-tenth of Penfield's experimental subjects being officially denoted as female, a specifically female homunculus was never published or known to be drawn by Cantlie. Nonetheless, in the late 1980s, a left female breast started to appear in the illustrations, usually in the motor homunculus, which Griggs traces back to an error in a 1979 *Scientific American* article by Geschwind that featured the homunculi (Griggs 1988, 105; illustration in Geschwind 1979, 182). Griggs playfully suggests that "Mrs. Cantlie has ghostdrawn these homunculi with breasts in order to redress the sexual bias prevalent in her earlier drawings" (1988, 106). Yet the issue of the absence of female body parts, or a particularly female version of the homunculus, is real, given the presence of female subjects documented among Penfield and his team's cohort. Among an estimated total of 1,065 patients over the course of Penfield's career, 107 were women, so just 10 percent, and 121 were left unrecorded (Di Noto et al. 2013, 1005). For Penfield and Rasmussen's 1950 mono-graph alone, 400 operations were recorded (Penfield and Rasmussen 1950, x) of which only 9 were specifically on women, equating to 2.25 per-cent. As a summary of research conducted over two decades, the book mentions women only a handful of times, and only then in regard to certain case studies of epileptic patients. These include the woman de-noted as "D.Sp" with the title "Foot and Hand Sensation" (111–13) and "E.C.," a 27-year-old woman (26–27).

Second, compensating for the historical paucity of data, the ques-tion of what a *her*munculus might look like can start to be addressed in the wake of recent studies. Penfield and Rasmussen's case study "E.C." is notable because she is the only subject of 107 to report any genital stimulation following their usual method of direct electrical stimula-tion of the cortex (Di Noto 2013, 1006). Penfield had removed a tumor from her right precentral gyrus, after which E.C. was found to have an unusual cortical map. The genital sensations that resulted from cortical stimulation involved a tingling that shifted between left breast, labia, and buttock. From this single data point, Penfield and Rasmussen sug-gested that the external genitalia and breast were localized in an area associated with the foot in the precentral gyrus motor area. This seemed plausible in the wake of an earlier study by Erickson (1945). Surprisingly, as Di Noto et al. (2013) report, a study using somatosensory-evoked po-tentials by Truett Allison and her team (Allison et al. 1996) confirmed that the clitoris and perineum are indeed mapped to the medial wall of the somatosensory cortex, anterior to the foot area. A more detailed study of the female genitalia using functional magnetic resonance im-aging (fMRI) in 2011 by Barry Komisaruk at Rutgers and colleagues is "the first to provide a female-specific sensory map of the genitalia— which includes the clitoris, cervix, and vagina—on the homunculus," report Di Noto et al. (2013, 1008). But significantly, this female genital

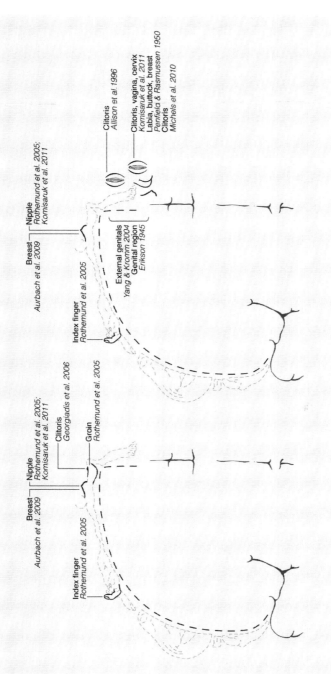

FIGURE 7. *Indicative diagram of a female homunculus ("hermunculus") mapped onto Penfield and Rasmussen's homunculus, a composite based on a range of recent studies. Illustration by Shelley Wall, coauthor of Di Noto et al. (2013, 1011). Reprinted with permission from Oxford University Press.*

Breast
Aurbach et al. 2009

Nipple
Rothemund et al. 2005;
Komisaruk et al. 2011

Clitoris
Georgiadis et al. 2006

Groin
Rothemund et al. 2005

Index finger
Rothemund et al. 2005

Breast
Aurbach et al. 2009

Nipple
Rothemund et al. 2005;
Komisaruk et al. 2011

Index finger
Rothemund et al. 2005

External genitals
Yang & Kromm 2004
Genital region
Erikson 1945

Clitoris
Allison et al. 1996

Clitoris, vagina, cervix
Komisaruk et al. 2011
Labia, buttock, breast
Penfield & Rasmussen 1950
Clitoris
Michels et al. 2010

somatotopy does not map homologously onto the placement of the cortical representation of the penis in males. In fact, Komisaruk et al. (2011, 6) confirm Penfield and Rasmussen's single data-point case study, "E.C.," who reported sensations arising within both the breast and the labia from one cortical region. But they also found additional sensory pathways between the cortex and internal female genitalia such as the cervix, differentiating them from male genital pathways, "which would suggest a different mapping of the genitals in the cortex" (Di Noto et al. 2013, 1009) (see Figure 7).

Because of this section's historical narrative I have concentrated first on cisgendered human subjects born with traditionally "female" genitalia who have been disregarded in the existing dataset. However, implications from recent research on cortical plasticity opens up possibilities for more dynamic somatotopic mapping, with more complex combinations of factors including changed gender identifications, fluctuations in hormone levels, ovulation, pregnancy, limb amputations, mastectomies, stages in the life cycle, and other alterations to the physicality of the body. For example, phantom sensations after mastectomy and the remapping of the sensory cortex in the wake of trauma are exciting indications of the place of plasticity in the formation of homunculi and hermunculi. In *The Brain's Body,* the sociologist Victoria Pitts-Taylor looked at ways that the concept of plasticity has been guardedly embraced by feminist empiricists as a way to critique the assumed fixity of brain, body, and identity within neurobiology, on the one hand, or the popular discourses, a "new master narrative of changing the brain-body" (2016, 18) demanded by technoscientific capitalism, on the other. Pitts-Taylor explains that, although feminists are highly critical of neuroscientific claims of bifurcated sex difference, studies of brain plasticity can be used in fact as a resource to critique sex/gender bias and as an alternative explanation for the evidence of sex differences found in the brain. This opens the gateway to seeing how "social forces, rather than evolutionary ones, are the cause of neurobiological difference" (Pitts-Taylor 2016, 23). Furthermore, although Pitts-Taylor does not mention the homunculus, the sexual difference at the heart of Paula Di Noto and her team's work on the *her*munculus would surely be allied.

CHAPTER 2

Pain as a Distinct Sensation
*From Psychophysics to
Affective-Motivational Pathways*

The pioneering research in psychology laboratories that, first, separated out pain from other sensations, and second, brought in standardized measures applicable across human research subjects through the
use of specific instruments for this purpose, initially arose in Germany
through the work of Ernst Heinrich Weber, and was continued by his
student Gustav Theodore Fechner. This chapter offers highlights of laboratory research as a way to foreground the shift in discourses about
pain and sensation that occurs in the transition from a generalized
"common sense" *(Gemeingefühl)* to a more spatially specific cartography of the body in the nineteenth century. Commonly, the perception
of changes in one's own body such as pain, fatigue, and movement were
initially scientifically understood as forms of, or variations of, touch.
Weber subjected such bodily sensations to scientific scrutiny, first in *De
Tactu* (On the difference in tactile sensibility) in 1834, and then in *Der
Tastsinn und Gemeingefühl* (On the sense of touch and common sensibility) in 1846, which put broader somatic sensations onto an empirical
footing. Weber's meticulous observations would inspire a new direction
in empirical psychology, and Fechner advanced the project through a
branch of psychology concerned with the precise measurement of sensation that he termed "psychophysics" in his *Elemente der Psychophysik
(Elements of Psychophysics)* of 1860. In large part, this chapter is the
story of laboratory-based measurements of sensation from the 1830s
onward. It treats the search for pain as a distinct sensation, and the
subsequent need to standardize its measurement, as a consequence
in part of the rise of the cultures and sciences of measurement more
generally, which led to innovations in the production of "normal" and
"deviant" subjects in criminology, such as in Cesare Lombroso's *L'uomo
delinquente (Criminal Man)* of 1878. But clearly the rise of measurement
and standardization was crucial to the progress of biomedical science,

where innovations in pain treatment through pharmaceutical developments in analgesics and anesthesiology advanced during that same period. For example, codeine was first isolated in 1832 in France by Pierre Robiquet, the first use of anesthesia for surgery was by William T. G. Morton in 1846 at the Massachusetts General Hospital, and morphine became more widely used for pain relief through intravenous injection in the 1850s (Porter 2006).

Before we proceed, however, the attentive reader might understandably balk at the suggestion that pain was indissociable from sensations either of touch or a generalized "common sensibility." Pain is unmistakably pain, one would think, and methods to moderate pain through the effects of plant-based naturally occurring analgesics have been known throughout human history, and across most cultures. On the one hand, this chapter focusses upon the rise of a particularly scientific approach that maps the sensory interiority of the body. The new empirical psychologists of the nineteenth century did not "discover" pain, of course, any more than Sir Isaac Newton "discovered" gravity. Their contribution was to devise consistent means to measure subjectively felt sensations of touch and pain and, through the invention and adaptation of suitable instruments, devise a mechanism for judging thresholds of sensation across a range of human experimental subjects. On the other hand, and seemingly counterintuitively, the isolation of pain and its recognition as a distinct phenomenon remains somewhat contested even in the present day. There is no mistaking the painful nature of the sensations involved after a freshly broken leg, for example. But a recent headline-grabbing psychology study by teams from the University of Virginia and Harvard University led by Tim Wilson in 2014 suggests that the experience of pain, and especially self-inflicted pain, is not so straightforward or distinct as we might imagine. By piecing together elements from the history of the physiology and neuroscience of pain and touch, this chapter sets up the groundwork to interpret this controversial laboratory-based study on pain as a stimulus. Some continuity is therefore established between Wilson et al.'s recent study and prior historical assumptions about pain, sensation, and its measurement that Weber, Fechner, and others were determining, as they put psychology onto a firmer empirical footing. In questioning this long-held impetus to measure pain as a distinct sensation, therefore, this chapter returns us to some difficulties in the subjective conscious interpretation of the unique qualities of particular interior bodily states, and considers what aspects of these sensations are accessible to measuring or recording apparatus to this day.

Pain and Boredom: An Experiment

"If I were to choose between pain and nothing, I would choose pain," the protagonist in William Faulkner's *Wild Palms* muses (Hardcastle 1999, 121).[1] This was the choice apparently offered to participants in a notorious and widely reported social psychology study by a team from the University of Virginia and Harvard University led by Tim Wilson, published in the journal *Science*. The title was innocuous enough: "Just Think: The Challenges of the Disengaged Mind" (Wilson et al. 2014). Given the abundance of electronic devices and screens constantly surrounding us, the rationale of the study was to see the effect of removing them, and consequently of leaving research subjects alone with their thoughts for a substantial and specified time period. Sensationalist headlines in the BBC, *Wired,* and *Forbes* claimed that subjects would rather shock themselves to induce pain rather than experience boredom. This chapter uses the Wilson et al. study as an opportunity to historicize the distinct and successive ways that nervous pathways and affective and motor functions of pain have been diagrammed within the body, up to and including the neuroscientific model that forms the basis for Wilson's discussion. Significantly, the history of physiological investigations into pain and sensation offers the means to diagram pain and sensation beyond the boundaries of the individual body, to consider the broader milieu of the organism and its environment.

The chapter is structured around the following four, somewhat interrelated, claims. First, in terms of laboratory-based experimentation and auto-experimentation with pain, there is a long history of regarding pain and touch through introspective means, a practice that attained scientific credibility in the nineteenth century through such experimental physiologists as Claude Bernard and Charles-Édouard Brown-Séquard in France; Johannes Müller, Ernst Weber, Gustav Fechner, and Alfred Goldscheider in Germany; and Magnus Blix in Sweden, the latter two independently conducting similar experiments on the perceptual thresholds of touch and temperature to test Müller's theory of specific nerve energies (see, e.g., Norrsell, Finger, and Lajonchere 1999; Finger

2001, 150ff). Second, pain and other bodily sensations became "measurable" in the mid-nineteenth and early twentieth centuries through the use of instruments and technologies that first demarcate, then inscribe, an introspectively sensible difference or limen. Through the pioneers in psychophysics Weber, Fechner, and after, the wide variability of subjective pain experiences was no barrier to the desirability to establish correspondences across and between human subjects, thus it was the differences in felt intensities themselves that were measured, then rendered visible as objects of study. The story of Weber's and Fechner's "just-noticeable difference" (JND) is partly the familiar story of machines of observation that capture vital processes and transfigure them as visible traces or representations in modernity (see, e.g., Crary 1992; Danius 2002; Brain 2008, 2016). Third, taking the idea of measurable intensities forward, we briefly consider the nervous system as itself part of a larger homeostatic system of "affective-motivational information" (Hardcastle 1999, 103; Auvray et al. 2010). This is an intermediate step toward, fourth, considering how pain is simultaneously an intraorganismic sequence of responses that tips the organism as a whole from perception to action, but also a biopsychosocial phenomenon that operates beyond the organism.

Left alone, without screens or outward distractions, our inwardly directed thought is known in neuroscience literature as "default-mode processing" (e.g., Gusnard et al. 2001; Gusnard and Raichle 2001; Wilson et al. 2014) or "default-mode network" (DMN) activity (e.g., Greicius and Menon 2004). However, one variation of Wilson et al.'s study in particular was responsible for the wider reporting and social media proliferation of the story. An electric shock generator was placed in the room, and participants were given the option to press a button to self-administer pain. Would participants prefer "noxious" stimulation to no stimulation? A little sociological context might help: previously, 83 percent of American adults claimed they spent no time whatsoever "relaxing or thinking" (U.S. Bureau of Labor Statistics 2012). One of the Wilson team's foundational research questions addressed whether this was because adults find "just thinking" unpleasant or aversive. Was "just thinking" aversive enough that subjects would willingly self-inflict pain? The team's means of assessing participants' affective states and their motivations for self-inflicting pain was down to self-reporting by participants after each session, and some problematic assumptions within their methodology are examined later. However, rather than routing the analysis directly within phenomenological analysis or cultural responses to pain (e.g., Bendelow and Williams 1995), or the embodied experience of those suffering and reporting pain (e.g., Honkasalo 1998), this chapter takes a different direction. Broadly sharing both of those papers' critical medico-psychological concept of pain, I take Wilson's headline-grabbing yet straightforward interpretation of pain as an alleviation from default-mode processing as a starting point to consider the quantification and administering of pain via nineteenth- and early twentieth-century experimentation and auto-experimentation. By means of early experimental physiology and psychophysics, and the use of a series of specialized instruments, I show how pain emerges as a distinct sensation and consequently a measurable object of study. Successive theories of pain start to widen the scope of pain beyond the individual organism and its nervous system, and involve more complex responses to immediate environmental contexts. Historically, this widening culminates in the recent "biopsychosocial" model of pain. But what of the *experience* of pain, and motivations for the self-administering of pain in the case of Wilson et al.'s subjects? What theoretical resources support a model of pain and sensation that exceed the boundaries of the individual body?

Elaine Scarry demonstrates that, historically, a central problematic of pain is that its expression has been a function of the prevailing discourses of the body at the time, whether through theories of humors, or mechanism, or in the final chapter of *The Body in Pain,* the "electrical nervous system" (1985, 282ff). Scarry's project has undoubtedly been intensely productive in theorizing relationships between body, pain, and world, and the history of science and medicine certainly offers compelling narratives and discourses about pain. However, my argument about bodies, pain, and (auto)experimentation is neither a strictly chronological history of the science of pain research (e.g., Rey 1995, Finger 2001), nor a cultural history (e.g., Mososco 2012). And, whereas pop neuroscientific explanations tend to reduce complex experiential phenomena to mere brain scans and seemingly self-evident neuroimaging techniques, I follow Scarry's advice in avoiding "analogical verification,"—that is, the appeal to a supposedly "sheer material factualness" (1985, 14) in explaining complex phenomena such as pain solely in terms of, say, prevailing models in cognitive science. When it comes to Wilson's study especially, the biopsychosocial reality of pain reveals the limitations of computational analogies (e.g., nerve pathways, centralized signal processing). Although Scarry's book is rich in theoretical explorations of pain and violence, and offers a multitude of threads that could be revisited and developed in novel phenomenological directions, the last two sections of the chapter shift the focus from the measurement of neuroanatomically "normate," or typical, human subjects of study to another level of abstraction, the neurophysiological state of the organism as a whole in its stance or orientation to an immediate environment. Consequently, I consider another perspective that has long been present within the history of science, based on the role of pain and sensation in the organism as a homeostatic entity. This perspective encompasses the nineteenth century concept of the *milieu intérieur* from Claude Bernard, the early twentieth-century physiologists Charles Sherrington and Kurt Goldstein, and aggregates into another approach to pain explored by twenty-first century neuroscientist A. D. "Bud" Craig. Here, pain is less about circuits, electrical systems, or measurable signals, and more about thresholds of difference, the perception of intensity, and the motivation of the organism as a whole. In the light of this, Wilson et al.'s study of pain through autostimulation might be considered as a manifestation of the "aliveness" that results from what Scarry calls the "felt-experience of sentience" (1985, 285). In other words, it is part of the nature of default-mode processing that the roiling fizz and buzz of peripheral nerve states distributed throughout the body become dampened and homogenized to stable background states. One initial explanation for Wilson's results I consider below is that self-inflicted pain offers just such a self-reflexive form of sentience, a way of individuating the organism in relation to its environmental situation.

Pain and History: Medicine

For early psychology and physiology at least, the history of the concept of pain was inextricable from that of touch. It was only through sustained experimentation from the mid-nineteenth century onward, as will be shown, that pain becomes a distinct and measurable sensation in itself. Yet through the successive series of models of pain that arise in the wake of such experimentation, an underlying dynamic will persist in the recognition of pain, since it is unusually experienced as both "absolute" and as "relative," thought Xavier Bichat ([1800] 1815, 53). To elaborate on that dynamic in this section, consider how nineteenth-century models of thresholds and nervous pathways within the individual human subject translate into twentieth- and twenty-first century models that accommodate modulation and even suppression of pain sensations in some cases.

Erasmus Darwin anticipated what is known as the intensive theory of pain, which assumed no dedicated or specific nerves were involved in its production. For him, pain was simply the overintensive excitation of nerves of any kind, whether for optical, auditory, or tactile perception. Pain results, he said, "whenever the sensorial motions are stronger than usual. . . . A great excess of light . . . of pressure or distension . . . of heat . . . of cold produces pain" (in Dallenbach 1939, 333). But Blix in 1882 and Goldscheider in 1884 had both found separate spots on the skin, and therefore separate nerves, for warmth, cold, pressure, and pain. Their work was pivotal in moving scientific research away from the intensive theory and toward a "sensory theory of pain"—that is, that pain is a sensation mediated by specific nerves. Pain would therefore operate as a sense of its own, distinct from touch. Blix and Goldscheider's research was championed by clinicians of the time, including Silas Weir Mitchell in the United States and Henry Head in England. The theory was verified in Head's published clinical observations, and in Germany with Maximilian von Frey's experimental results on skin "correlating warmth with Ruffini cylinders, cold with Krause end-bulbs, pressure with hair follicles and Meissner corpuscles, and pain with free nerve endings" (Dallenbach 1939, 340). Unlike those other sensations, however, pain was held to be averse to adaptation, the one sense that defied habituation. It was only in 1919 that experiments at Cornell by Straus and Uhlmann finally showed that pain acts like any other stimulus, such that organisms adapt to a relatively constant and unvarying form of pain. The reason pain was previously considered resilient to habituation was because of the "constantly varying" nature of "stimulus conditions" of pain in the cases of injury, toothaches, headaches, and so on (343). Any rhythm or periodicity to cycles of pain, in other words, furthers its felt constancy. If a distinct sensation of pain relies on a just-noticeable difference [JND] from a painless base state, successive cycles of pain nerve stimulation result in a *very* noticeable difference in the internal body state of the organism. This could form one explanation for

Wilson's study, that discrete painful shocks were self-administered in order to induce a felt intensity to stand out from a normatively painless, stimulus-stable somatic background state. This would be pain as reappropriation (where *proprius* is literally "one's own"), a fully intentional rediscovery of somatic physicality through an interoceptively perceived bodily state.

Over this and the following section, I show how the value of Wilson's controversial study lies in the way it reexamines thresholds of perceived pain only in its relation to the ongoing outward-directedness of both autonomic and conscious processes of the organism. In other words, Wilson's study serves as a recent waypoint on the historical path to regarding the disposition of the nervous system toward outward excitation, and the autoproduction of intensities, in the organism as a form of self-regulation between its inner and outer environment. At least in primates and humans, such self-regulation implies both motoric and affective responses, simultaneously straddling both preconscious autonomic processes and conscious states and motivations. The gestalt neurologist Kurt Goldstein's *The Organism* (1934) specifically conceptualizes the relationship between the organism as a whole, involving individual reflexes, organs, and pathways within the interior of the organism, what Bernard (1865) had called the *milieu intérieur*. My historical retracing will also encompass a more recent neurophysiological framework from A. D. Craig for considering pain as homeostatic emotion, involving affect and motoric drive. This strand of thinking about pain departs from prior neurophysiological explanations that sought to explain even complex and heavily contextual pain reactions by means of specific reflex responses and nervous subsystems. For example, the neurophysiologist Charles S. Sherrington (1906), revisited in more detail below, assumed an underlying stimulus-response mechanism involving pain-perceiving "noci-ceptors" in the body that fed into sequential reflex loops. This organizational logic is evidently recognizable in Melzack and Wall's "gate control theory" of pain, based on then-contemporary analogues with electronic circuitry (Melzack and Wall 1965). Patrick Wall had seen that Sherrington and his pain research contemporaries were focused on "a physiological event" rather than "anything to do with behaviour and sensation" (Wall, interviewed in Rosenberg and McMahon 2004, 77). When Wall moved to MIT to work with Norbert Wiener in

1953, he also met Melzack, and carried forward the cybernetic model in their classic 1965 *Science* paper.

In their paper, Melzack and Wall distinguish between two models of pain, and propose a third. First, the "specificity theory" assumes pain is a modality, like any other sensation. Descartes had proposed in his 1644 treatise *L'Homme* that there is a direct pathway to the brain, like ringing a bell at the end of a rope (1965, 972), for example. Second, developing Goldscheider's model, "pattern theory" proposes that, in addition to the intensity of a stimulus like the ringing of a bell, there is a spatio-temporal pattern of impulses coming from various points of the body that are then assessed through central spinal cells, a mechanism they term "central summation" (973). This allows variability in afferent nerve input, but as a result of insufficient neuroanatomical knowledge of the time, Weddell vaguely hypothesizes in 1955 that the central processing occurs within spinal cord activity. Melzack and Wall proposed a third model based on new physiological evidence from spinal mechanisms—namely, experiments involving a strand of tissue generically termed *substantia gelatinosa* that stretches the length of the dorsal column (now termed the posterior column-medial lemniscus, or PCML). Like the pattern theory, their model allowed spatiotemporal variations of afferent input, but like a gate control in electronics, a modulation process allowed only certain relays from the spinal cord up to pain processing areas of the brain. It was illustrated in their paper by means of a circuit diagram, with peripheral inputs on one side, an "action system" (output) on the other, and a "central control," the gate that modulates the signals, in the center (975). Unlike previous models, this allowed for the inhibition or dampening of pain signals, a well-recognized psychological phenomenon in phantom limbs or in high-stress situations, such as sports games or the battlefield, where subjective pain sensations are absent and are actively suppressed. Or consider the rare neurological condition of pain asymbolia, where pain is experienced without accompanying unpleasant sensation, which Grahek terms "painless pain" (2014, 95ff). Here was a plausible physiological explanatory model for the complex psychology of pain, and so rather than "the ringing of a bell" (Descartes's model), or "a sensory adjunct of an imperative protective reflex" (Sherrington's model), it was instead "an ongoing process" (Melzack and Wall 1965, 976).

Pain as an ongoing process is a step in the right direction for a wider biopsychosocial conception of pain. Melzack's research continued to place the study of pain on a firm neurophysiological footing. In 1971, Melzack and Torgerson developed the McGill Pain Questionnaire, which sought to categorize different sensations of pain by associating them with particular groups of words, some of which denoted the subject's emotional state (Melzack and Torgerson 1971). Later still, Melzack developed the "neuromatrix" theory with regard to phantom limb pain (see, e.g., Melzack 1990). Explanation of embodied processes in terms of circuits, and later matrices, had much currency during an era in thrall to the social as well as the technological possibilities of cybernetic theories (see, e.g., Bateson 1972). However, as a model it fails to adequately explain the "social" aspect of pain, something at the heart of a recent wave of research that interprets pain as biopsychosocial (see, e.g., Gatchel et al. 2007). Thus, while gate control theory is a historically significant attempt to combine the psychological and the physiological elements, I look elsewhere to widen the explanatory reach beyond individualized somatosensation, to incorporate (more of) the "social." Rather, a homeostatic view of pain seems consistent with Goldstein's conception of the response of the organism as a whole to its environment, where pain from intense heat or cold, for example, only becomes a prompt for action or physiological change as a result of the organism's affective or self-regulatory response to relative conditions between it and its environment. The neuroscientist "Bud" Craig summarizes this new scope of the concept of pain:

> Pain is an enigmatic feeling from the body, distinct from the classical senses because it is multifaceted (it is a discriminative sensation, an affective motivation, a potent autonomic drive and a reflexive motor stimulus) and because it is inherently variable. (2003, 303)

In the same paper, what Craig terms an "homeostatic emotion" is an attention-demanding sensation and motivation that derives from an interoceptively perceived bodily state (e.g., thirst, hunger, fatigue) that drives behavior (respectively: drinking, eating, and resting) aimed at maintaining the body's internal state (Bernard's *milieu intérieur*). As will be discussed below, pain is one of a number of interoceptive states yet it also lies at the interface of autonomic bodily processes, affective drives, and conscious control. If, according to Craig, research findings "indicate that pain in humans is a homeostatic emotion reflecting an adverse condition in the body that requires a behavioral response" (307), then our question becomes: Can human or nonhuman subjects be so averse to boredom or lack of external stimulus that this constitutes an adverse condition in the body, and so prompts activity or self-produced stimulation to the point of harm?

PAIN AND BOREDOM: WHAT DOES WILSON'S STUDY TELL US?

There were 11 variations within Wilson's boredom study. Subjects were kept alone in an empty room without distraction, either in a laboratory or their home, unable to use objects including mobile phones or writing materials. The time periods varied, being either 6, 12, or 15 minutes, after which subjects were required to report, first, their level of enjoyment and, second, any difficulties in concentration. In the first phase only university students were recruited, and later subjects were recruited from a farmers' market and a local church, spreading the age range from 18 to 77. Questions can be raised about methodology and sample size, since the researchers claimed they factored out age, education, income level, and the frequency with which subjects used smartphones or social media, and the total number of participants being only 42 (Wilson et al. 2014, 76). But this would be to miss the point that the media gleefully alighted upon, with attention-grabbing headlines such as "Men would rather receive an electric shock than think" in *Wired* (Clark 2014), and "Students prefer jolt of pain to being made to sit and think" in the *Guardian* (see, e.g., Sample 2014). For Wilson's team did indeed incorporate a modification of their experiment for some of the 15-minute sessions. For a certain set of participants, all of whom had been identified as being so pain-averse that they would pay the researchers not to experience the negative stimuli of an electric shock, they now had the option in the distraction-free study time to self-administer electric shocks on their ankle by pressing a button. "We went to some length to explain that the primary goal was to entertain themselves with their thoughts and that the decision to receive a shock was entirely up to them," justified the authors (2014, 76).

As presented in their paper, the statistics seem to unambiguously support their narrative. 67 percent of men (12 of 18) gave themselves at least one shock during the thinking period, the mean being 1.47 shocks per person. For women, 25 percent administered at least one shock, the mean being 1.00. Interestingly, among the men, there was one outlier who shocked himself 190 times (76), which, although noted in the article and in the media, nevertheless remained unexplained, and was presumably discounted from the average. The BBC News coverage, headlined "Do people choose pain over boredom?" was typical of most media in focusing on the painful shock aspect:

> Prof Wilson's team did the electric shock experiment to try to find out if quiet, solo thinking was unpleasant enough that people would actually prefer something nasty to happen. Sure enough, 18 of 42 people, more of them men than women, chose to give themselves at least one mild shock on the ankle when left alone for 15 minutes. (Webb 2014)

Of course, at a period when the properties of electricity were still being charted, an explosion of inventiveness in the nineteenth century offered many different means of self-administering electrical shock as therapy. As Carolyn Thomas de la Peña writes, a utopian vision of the possibilities of electrotherapy was prompting new applications. Electricity was proven to work in relieving pain and treating partial paralysis, for example. "These, however, were only a few of the ailments that physicians 'cured' with electrotherapy," she notes. "By the 1880s, electrotherapy manuals held that physicians could cure everything from poor eyesight to sexual dysfunction with a small battery-powered device" (2003, 95). During this period a picture is built up whereby inventors, entrepreneurs, and even the interested public were able to experiment with electricity, either on themselves, on others, or both. Laboratory-based autoexperimentation regarding pain and the sense of touch had occurred earlier in that century (described in the following section). However, these historical remarks concerning electricity are offered in the light of the sensationalized reporting of Wilson's experiment, which crushes a prior discourse of curiosity through self-experimentation in favor of a simplistic binary of "boredom" and "stimulation." Furthermore, is the likelihood of a painful outcome through self-experimentation really to be contrasted with sitting and thinking by oneself?

The editorial perspective on Wilson's series of experiments within the BBC article rather predictably fit into popular narratives of the downsides of excessive screen time, which is associated with negative social effects such as distraction, disengagement, and an unhealthy need for constant stimulation: "People are unhappy in their own company and some prefer painful experiences to their own thoughts" (Webb 2014). However, upon further inspection of the dataset, a commentary paper by Fox et al. (2014) demonstrated how Wilson's conclusions were false. Pointing to inconsistencies in the data, where "just thinking" was not actually consistently characterized by participants in negative terms, and also to an increasing body of research in the affective qualities of thinking and daydreaming, Fox et al. instead argued: "It is impossible to draw meaningful conclusions regarding the 'typical' affective qualities of spontaneous thought, given their enormous variability both within and across individuals" (1). Whatever commonsense assumption we or the media might make, "just thinking" is hardly a consistently experienced qualitative phenomenon. Furthermore, as Valerie Hardcastle writes, philosophers often misconceive the experience of "pain" as "an intuitive and unproblematic example of conscious experience" (1999, 96). Historically speaking, before pain became a scientifically measurable and distinct sensation, it was long considered to be associated with touch in various ways.

Pain and Touch: Measurement

Wilson's study offers a useful opportunity to consider how we perceive and monitor internal bodily states against a background of habituation to constant stimuli. The perception of our own bodily states has had a complex and nonlinear history, with continued confusion as to its relationship with the normative schema of the five sense modalities. After Aristotle and the Early Modern period, this form of inwardly directed perception was broadly recognized, but given many different names, including "inner touch" (e.g., Heller-Roazen 2007; Paterson 2015). Regarding such subjective sensations, Mososco identifies the shift: "As opposed to introspection and testimony, the new science of the intimate sense had to be rooted in physiology and physics" (2012, 106). Yet, for the forerunner of psychophysics Ernst Heinrich Weber at least, the measurable sense of touch *(Tastsinn)* remained related to *Gemeingefühl* or "common sensibility" in his 1851 study (Weber 1996). Pain would become identified and subsequently quantified as sensation. Weber was the first to measure tactile perception in the laboratory, albeit a small range of mostly adult male subjects, as both women and children he considered too unreliable in their reporting of subjective states (1978, 54). This remained a step toward the rationalization of the senses, and specifically the idea of a neuroanatomically normate human subject with respect to tactility. Fechner, who built upon Weber's experiments and became known as the "father" of psychophysics, stated that the aim of his project was no less than to establish "the exact science of functional relations or relations of dependence between the body and the mind" (Fechner [1860] 1966, 7), and aimed to map the neuroanatomically normate (adult, male) human also in terms of pain. The measurement of touch and pain would seem to be situated exactly at this interface. Yet, while initially deemed experientially related to touch, the measurement of pain would offer its own challenges within the laboratory by virtue of the limits of the identification of felt intensities, and the adaptability and habituation of the body as a whole to pain stimuli.

In practical terms, if vision and audition are measurable through direct methods, pain and touch necessitated the development of indirect procedures. Certain bodily sensations such as pain, or the feeling of moving a weighted object, could not be directly measured through a consistent and invariable stimulus (e.g., light intensity for vision, audible tone for audition) across experimental subjects. It would have to depend on the relative *difference* between default or background bodily states and newly induced subjectively experienced sensations. This culminated in what Weber called the "just-noticeable difference" (JND), also known sometimes as the "differential threshold" or "difference limen." It was later refined by the experimental psychologist Fechner, who maintained that it held true across the range of sensory capacities, including pain but excepting smell, in his *Elemente der Psychophysik* of 1860. The need for a standard unit of measure of sensation and their associated laws, applicable across human experimental subjects, was premised on the idea that sensations would be noticed more precisely, and therefore stand out in terms of difference and intensity, within certain bodily areas rather than others. The resulting "map" would place denser concentrations of nerves in more sensitive skin areas, and diffuse nerves in less sensitive areas and the viscera. As Weber expressed it in *Der Tastsinn* of 1846, this entailed a conscious attention to sensation from the outside, from the bodily surface, as opposed to the bodily interior:

Our sense-organs are directed outwardly not inwardly, in order that the mind may receive impressions from the external world; it would become very confused if internal processes were persistently demanding its attention. One intestinal canal touches and rubs against another, lungs rub against the skin of the pleura covering the chest cavity, muscles press and rub against each other: but we have no sensations of these. (Weber 1996, 148)

Thus Weber, followed by Fechner, Goldscheider, and others, were effectively embarking upon an experimental cartography of touch, pain, and the nervous system by mapping out the sensitivity of the exterior of the body, of intensities and areas of cutaneous difference, by means of an invariant—or, at least, precisely variable—instrument. Inevitably, some differences would be more "noticeable" than others, and some locations on the cutaneous surface involved more touch and pain discrimination than others. Weber had already conducted extensive experiments on perceptions of touch in human subjects in his 1834 book *De Tactu* (On the difference in tactile sensibility) and then *Der Tastsinn und Gemeingefühl* (On the sense of touch and common sensibility) (1846), and his work strongly influenced Fechner. If experimental subjects could discriminate between, say, touch, pain, or itching beneath the skin, his reasoning went, then the necessary nervous structure that supported such sensations could be discovered. David Parisi (2011) makes the case for what he calls the "rationalization of touch" and its disassembly into component parts via laboratory experiment in the nineteenth century. His argument follows the development of concepts—as well as the instruments—of measurement of tactile stimuli from Fechner, Wundt, and Weber. Weber in particular he regards as a "foundational figure" because his "method implicates touch in a discourse of modernization" (2011, 196), a discourse that ordinarily neglects that modality. The "two-point threshold" or "two-point limen" for the discrimination of touch that Weber advanced employed compass points to detect when two touch sensations on the experimental subject's skin were felt as one. He designed and used such a precisely variable instrument, describing it in *De Tactu* as a *Stangenzirkel* (beam compasses) (1996, 30), a handheld metal frame with a printed scale with two points, one that could slide closer or farther from the other. Later devices would be larger, and some affixed to tabletops.

Since the threshold marked the minimum distance on the skin that two distinct points resolve into a single sensation of touch, different areas of the subject's skin inevitably have different sensitivities and therefore produced widely varying results. As Parisi explains: "The two-point threshold experiments produced a map of the skin as a surface with varied acuity for discriminating between stimuli" as, at the tip of the tongue, "subjects could perceive the compass points as distinct when separated by only 1.12 millimeters" (2011, 190). The chest had far less acuity, detecting the two points when spaced 45mm apart. Weber's search for a standardized measurable fidelity of tactile discrimination across adult male subjects led to what Parisi calls a "normative psychoanatomical model of the tactile system" (199). However, no matter how precisely Weber sought to map and measure touch sensations, they remained compromised, indissociable from background sensations of a "common sensibility," the sense of the body's perception of itself *(Gemeingefühl)*. This common sensibility for Weber is comparable to Craig's formulation of interoception, as it involves "the ability to perceive our own sensory states" (Weber 1996, 213). For example, there was motor activity of the limbs, and therefore *Muskelsinn* (muscle sense, as we saw in chapter 1); but also temperature (thermoreception), which affected skin receptivity; and of course pain also hindered the discriminatory capacity of a rationalized touch, adding emotional reactions to supposedly pure sensation, and inevitably imparting noise into any putatively measurable signal. Furthermore, as part of the common sensibility, sensations of pain were not as capable of variation or difference as the other, outwardly directed senses, thought Weber: "So much is certain, that qualitative differences between pains and other feelings of general sensation are much less numerous than feelings from the special senses" (1996, 153).

Compared with the methodical capture of discerned cutaneous sensation through instruments of touch, other sensations of the bodily interior including pain proved more difficult to map. Weber and Fechner, so careful about measuring stimuli on other human subjects, were not above practices of self-experimentation. In experiments on afterimages, Fechner famously stared into the sun for hours at a time: "I ruined my eyesight," he wrote, "looking often at the sun . . . so that by Christmas 1839 I could no longer use my eyes" (in Solomon 2011, 1).

Ernst Weber tried self-experimentation with pain, including hitting his thumb with a hammer and timing the interval between the initial sensation of pressure and the ensuing throbbing. His brother Eduard also self-experimented with electric shock by filling his ears with water, placing metal rods in them, and "evoking, through the induction of two very large magnets, a strong current" (Weber 1996, 165). Through such experimentation and auto-experimentation, says Mososco, "harmful experiences allowed the establishing of functional correspondences between injuries and sensorial intensities" (2012, 106). A small prick of the thumb, compared with a full hammer blow, might reveal different sensations, but also potentially different nervous pathways. Since Weber's investigations predated any reliable measure of pain such as Von Frey's aesthesiometer, then, he regarded the phenomenon of pain as compromising the measure of "pure" or rationalized touch. For Weber, pain could only impede this newly rationalized subject of tactile experimentation, and so became associated with the prerational, the instinctive, the animalistic:

> One effect of the intensity of many feelings provided by common sensibility is that the mind is prevented from calmly contemplating them in the manner necessary for the sensation to be referable to objects. Instead, the attention of the mind is driven by the pains to its own state of suffering, and its own body. The effect of this is that the sensations excite not so much the cognitive faculties as the faculties of desire, so that *we are driven to avoid pain by instinctive or intentional movements.* (Weber 1996, 153, emphasis added)

Appropriately enough, the nature of pain here is likened to an irrational and aversive force that compels the organism, either consciously or preconsciously, to avoid that sensation. That is, it highlights some of what Fechner considered the "functional relations" or "relations of dependence" between body and mind, a relationship that frames (the apprehension of) a stimulus and the potential motility of the organism. I return to this idea below.

The measure of perceptual thresholds became known as the Weber–Fechner Law, which in its most basic form stated that the intensity of the perception of a sensation is proportional to the stimulus that causes it.

By means of such indirect measurement, a mathematically determined perceptual threshold is produced based on what a number of human subjects notice, such as an increase in the intensity of light, or a difference in skin pressure, temperature, or weight. "Objective sensations, such as sensations of light and sound, are those that can be referred to the presence of a source external to the sensory organ," explained Fechner, whereas "changes of the common sensations, such as pain, pleasure, hunger, and thirst, can, however, be felt only as conditions of our own bodies" ([1860] 1996, 15). This architecture of measurement based on noticeable difference was employed in some form into the mid-twentieth century. Von Frey's experiments between 1894 and 1896, for example, used an "aesthesiometer," which involved a stimulus made from a strand of hair within an adjustable handheld instrument that could be varied precisely in terms of length or weight. The precision and constancy of the equipment was to eliminate variations in the stimulus across different subjects, and between related experiments on the same experimental subjects in order to identify and measure the intensity of touch, heat, cold, and pain at different points on the skin (Rey 1995, 234).

But if touch intensity could be mapped with instruments, what exactly was being measured in the case of pain? The pressure algometer, an instrument for measuring pressure or pain with a needle point, was functionally comparable to Von Frey's aesthesiometer, as both were to pinpoint the threshold whereby an experimental subject begins to discern pain as a sensation. Compare the measurement of one human subject with another, and the difference in felt intensities becomes significant. Algometers and aesthesiometers alike were designed to "standardise and quantify the intensity of a pain-producing stimulus" and "to map sensitivity on body surfaces, to study biosocial difference and to measure analgesic effect," explains Tousignant (2014, 112–13). The measurement of biosocial difference was indeed central to pioneering criminologist Cesare Lombroso, whose algometer was adapted from Emil Du Bois-Reymond's design. It consisted of an induction coil and a bichromate battery, and in *L'uomo delinquente (Criminal Man)* (1878), Lombroso showed it was useful not only in determining pain thresholds, but also in quantifying the insensitivity of body parts. The hands, throat, breasts, tongue, and clitoris of "normal" and "criminal" women were subjected to the induced current, and "the 'greatest deadness' was found in the hands of peasant women and in the clitorises of pros-titutes" (Lombroso, in Horn 1995, 116). Passing the algometer over vari-ous surfaces of the body, Lombroso explains how "the strength of the induced current is increased until the patient feels a prickling sensation in the skin (general sensibility) and subsequently a sharp pain (sensi-bility to pain)" (in Lombroso-Ferrero [1911] 1972, 246–47). Clearly, pain was regarded as a more intense stimulation of what Weber had termed the common sensibility. But Lombroso and Weber were poles apart in terms of the need to demarcate a neuroanatomically "normal" as op-posed to a neuroanatomically "deviant" subject.

Blix in 1882, and Goldscheider in 1884, following the lead of Müller and his theory of specific nerve energies, also helped to experimentally differentiate pain sensation from this generalized common sensibility or somatic sensation. From the late nineteenth-century and into the mid-twentieth-century, introspective methods were used to generate normative neuroanatomical models of pain, and the predominant mode of charting what is to be counted as distinctly experienceable, and indeed expressible, in terms of body, pain, and sensation using Weber–Fechner methods for decades afterward. For example, Hardy, Wolff, and Goodell at Cornell University in 1940 established an arbitrary pain intensity unit called the *jnd* (after Weber's and Fechner's acronym) that scaled from a barely detectible prick on the skin to a full throbbing pain. In what became known as the Hardy-Wolff-Goodell scale, there were two *jnds* to each *dol* (based on the Latin word for pain, *dolor*), and so the equipment manufactured to measure these pain units would inevitably be known as a "dolorimeter" (Mososco 2012, 107). Wolff was seen in fact as the "father" of psychosomatic medicine, linking pain in Tousignant's words to "emotional events and predispositions" (2014, 116), yet simultaneously wishing to bypass the emotional content and the vague expressions of pain in order to gauge pain "as a discrete sen-sation" (Wolff, in Tousignant 2014, 116). As the aesthesiometer did for touch, the dolorimeter worked as a machine for producing constants out of variability in pain responses. Furthermore, emotional differences across persons complicated the experience of pain, and so precision de-pended on depersonalized sensations of pain—that is, pain minus the inconstant variable of emotion. The Hardy-Wolff-Goodell scale was dis-credited in 1957 by the distinguished clinical anesthesiologist Henry K.

Beecher, who became chairman of the Committee on Drug Addiction and Narcotics and who established a Laboratory of Anaesthesia (Tousignant 2014, 122).

Pain and the Nervous System: Mapping Speeds and Intensities

Until halfway through the nineteenth century, arguably through to the work of J. H. S. Rease in 1848, in fact, pain had yet to be fully dissociated from the sense of touch (Rey 1995, 213). A crucial intermediate stage was Charles Sherrington's discovery of "nociceptors," a receptor specifically responsible for pain detection that signals through peripheral nerves. Knowing there was no single organ responsible for pain, Sherrington published a 1903 paper claiming evidence for peripheral nerve endings capable of detecting injury, and therefore responsible for signaling pain, a subsystem he termed "nocicipient" (1903, 40). In *The Integrative Action of the Nervous System* (1906), he expanded his concept of the "noxious stimulus" to include motor outcomes, one "with an intensity and quality sufficient to trigger reflex withdrawal, autonomic responses, and pain, collectively constituting what he called the nociceptive reaction," explain Woolf and Ma (2007, 353).[2] If scientific exploration into the mechanisms of pain have no single organ to investigate, nor even a specialized nerve pathway, the specialized nociceptors that Sherrington did find turned out to have an even more diffuse neurophysiology. There could no longer be any direct one-to-one signaling between receptor and brain area when harmful stimuli were detected. Increasingly, pain would have to rely on contextual input, processed and either amplified or dampened according to the organism's environmental situation. Sherrington's (1906) discovery of distinct pain receptors ("noci-ceptors") and specialized nerves in the twentieth century would help to dispel that assumption, although Sherrington explained reactions to such noxious stimuli through nociceptors purely in terms of systems of reflexes.

An explanation at the level of the organism would later be developed by Kurt Goldstein, influenced by the so-called father of modern experimental physiology, Claude Bernard, who first articulated the organizing forces that regulate the balance of chemicals and processes, both within the organism *(milieu intérieur)* and in relation to its environment *(milieu extérieur)*. While space does not allow a more detailed exploration of the concept, Bernard explains the role of the nervous system in regulating the *intérieur,* and the potential role of pain for an organism:

> The necessary conditions for the life of the elements which must be brought together and kept up constantly in the *milieu intérieur* if freedom and independence of existence are to be maintained are already known to us: water, oxygen, heat and reserve chemical substances. . . . These are the same conditions as are necessary for life in simple organisms: but in the perfected animal whose existence is independent, the nervous system is called upon to regulate the harmony which exists between all these conditions. (Bernard, in Hirst 2010, 41)

The neuroanatomy of pain has been reasonably well understood since Head's research from 1898, published in 1920, and Sherrington's research on "noci-ceptors," the sensory nerve cells that detect cell damage, discussed in his 1904 Silliman Lectures at Yale, which were published as *The Integrative Action of the Nervous System* in 1906. A brief overview of current neuroanatomical understandings of pain will suffice here. In vertebrates at least, the sensory subsystems dealing with pain take feedback from such "nociceptors," distributed around the periphery of the body, and route it to the central nervous system by way of the "dorsal horn" (one of the columns of the spinal cord) either to the "reticular formation" (a part of the brain stem with interconnected nuclei) to the thalamus, or else from the thalamus right up to the frontal cortex or the somatosensory cortex (see, e.g., Hardcastle 1999, 102). The different routes are taken according to different nerve fibers that inflect the character of pain, either dull (slow), or sharp (fast). The thalamus receives many types of sensory signals, and both types of pain signals. The faster (sharp) pain signals travel along $A\delta$ nerve fibers, which are myelinated, meaning sheathed and so more conductive, and end up in the somatosensory system, an area that deals with the planning of, and capacity for, movement. But the slower (dull) pain signals travel in thinner, unmyelinated C-fibers from the thalamus to the frontal cortex, and the pain experience in this case is accompanied by affective content, or as Hardcastle puts it, the reticular formation processes "affective-motivational information" (103) that places pain within the "perceptual-emotional" system (118). This neuroanatomical "snapshot" is offered at this stage to underline the shift in discourse about pain as sensation that occurs at the end of the nineteenth century, where pain had been almost inextricably associated with the modality of touch, and reveal an alternative trajectory from the early twentieth century onward, where a differentiated nervous cartography of pain became associated with affects and the perception of difference. How this framework aids the interpretation of Wilson's 2014 study is considered below.

Pain and the Organism: Movement

Rather than mapping the interior of the body and its nervous pathways as a way to understand pain, or conceiving of human and nonhuman organisms like Cazeaux, as "mere physiological beings," the gestalt neurologist Kurt Goldstein offers the perspective of the disposition of the organism as a whole in its response to external stimuli. Pain is clearly one such external stimuli, arising directly from an organism's interactions with its environment, and yet setting in chain a series of reflexes and neurophysiological responses in the *milieu intérieur* of the organism. It is surprising that pain does not feature more as a thematic. From Sherrington's discovery and analysis of reflexes in the nervous system of animals he found that small variations in the stimulus produce surprisingly varied responses in the organism, where a reflex may or may not be effected. Goldstein gives the example of a decerebrated cat that automatically swallows water if placed on the pharynx, the part of the throat behind the nasal cavity, but upon the addition of a small amount of alcohol the reflex is suppressed. Sherrington detailed a multitude of such observations and the variability he assumed was explicable in terms of a detailed mapping of the expression of different reflex responses to different kinds of stimulation. But Goldstein argued that this explanation was inadequate, that a reflex is not "a simple connection between a specific stimulus and a definite reaction" ([1934] 2000, 72), since there are different classifications of experience for an organism given the same stimuli: a stimulus could be regarded as "harmful" for the organism in one context, but "harmless" in another, for example.

In addition to being the first to identify and name "nociceptors," Sherrington conceptualized them as sensory neurons distributed around the body that respond to damage or deformation through feedback to the spinal cord. Hence conscious pain was registered with an associated movement by the organism, such as withdrawal of a limb from a painful encounter, the so-called flexor (or withdrawal) reflex. To be fair, in discussing reflexes resulting from stimulation to "noci-ceptive" apparatus, Sherrington (1906, 226) recognized how the experience of skin-pain was not so different in nature from the pain that resulted from excessive stimulus of other senses, or the intensity of sensations of heat or cold. But he did argue that the neural pathways were very different in each case. The stimulation of touch sensors (a "touch-spot") by mechanical pressure—a grip or a vice, say—in itself could never produce pain, as a different sensor ("noci-ceptor") is involved in order to produce the distinct pain sensation. So, while Sherrington acknowledged an experiential relationship between sensory stimuli such as touch and pain, where one might shade into another due to increased intensity, the neural pathways were entirely distinct, and the reflex responses therefore encoded separately. Furthermore, receptors from the same sensory system (such as noci-ceptors, say, or touch receptors), if they are situated close together, will "mutually reinforce reaction" (322) and hence heighten the intensity of the sensation. Yet, as noted before, "where members of two different systems lie close together, e.g. tango-receptor and noci-ceptor, in one and the same piece of skin, they . . . have conflicting mutual relation" (322), he said, so that one system (usually pain) prevails over the other.

Yet, for Sherrington as for others, the legacy persisted of bracketing pain and touch together, even if subserved by different neural pathways and therefore considered as differentiated systems. Experiments on frogs and dogs had suggested the existence of a "receptive field" for reflexes such as the scratch reflex, and in fact such reflexes are actually "*groups* of reflexes," using "approximately the same motor apparatus in approximately the same way" (129). This purported receptive field contains receptors not only for touch but also for somatosensation such as chemical, temperature, and pain stimuli. There is the potential, therefore, for both conflict and corroboration when different sensory subsystems are invoked through the reflexive field. For the previously mentioned flexor reflex, in the case of a dog's leg withdrawing immediately from pain, Sherrington argues: "Skin-pain is sensually incompatible with pure touch, the dolorous suppressing the tactual, just as the noci-ceptive reflex in the 'spinal' dog's hind leg suppresses the merely tango-receptive. But gustatory and tactual sensations excited from the same receptive surface, e.g. the tongue, habitually blend harmoniously"

(131). Since pain reflexes from noci-ceptors always trump tactile sensory input, the possibility of damage to the organism as a whole is minimized.

Goldstein similarly observed separate neurological functions of touch and pain in the organism. He observed: "The flexor reflex in the dog can be elicited through pricking, heat, pinching, and chemical stimulation, but not through touch and simple pressure. This means a differentiation between a more 'neutral' and a more 'unpleasant' character of a stimulus. Therefore, one can classify stimuli into 'harmful' and 'harmless' and assume that the effect will vary according to the harmfulness or harmlessness" ([1934] 2000, 72). As we know now, it is further complicated by the fact that nociform reflexes (i.e., reflexes occurring through stimulation of nociceptors) prevail upon other simultaneous sensory reflexes—hence the sense of pain becomes overpowering in a too-firm handshake, for example, even though the initial touch sensations which preceded it may have started off pleasantly enough. And there is yet another complication, since the reflex can be reversed: "In human beings, even in the face of pain and injury, no avoidance reflex will appear if the subject needs to obtain information regarding the nature of the stimulus" (72). This clearly implies a relationship between conscious motor control and preconscious reflex that is bidirectional, unlike Sherrington's straightforward neural model of pathways and circuits through the spinal cord.

Later, referencing Sherrington's withdrawing dog leg, Goldstein applies this argument about variability of responses, and the added complexity of conscious control, specifically to humans. When the chorus of reflex reactions stimulated by a pain stimulus is set in chain, carefully cutting the spinal cord reduces the "noise" and strips down the complexity to the most basic reflex:

> Stimulating the sole of the foot by a pinprick produces withdrawal of the leg. At the same time, however, pain is felt, and various corresponding phenomena appear in the whole of the body: the muscles, the vaso-motors, the pupils, and so on. If, however, the sensory tract to the cortex is interrupted through a lesion of the spinal cord, we may obtain only a reflex phenomenon without any of the other reactions— thus the reaction seems to be much simpler. (175)

Directly afterward, Goldstein discusses how, without such surgical intervention, the heterogenous distributed pathways of reflexes set in train by an external stimuli lead, at the level of the whole organism, to homologous effects. Furthermore, despite (or possibly as a result of) these multiple, parallel, heterogenous pathways, the previously outlined "reflex reversal" can still be accommodated in the organism:

> On the other hand, however, the effect in the leg can be prevented by conscious effort, and the concomitant reactions become more intense. The dependence of the distance of spread of the stimulus effect on the condition of the whole organism (which can be codetermined by still other simultaneous activities) shows itself most clearly in case of so-called reflex variation, in which entirely different systems are suddenly activated by the same stimulus. (175)

This leads to a conception of the organism as a whole that simultaneously involves the potential for a multiplicity of responses due to variability along the same (but parallel) neural pathways. If other factors are helpfully and literally cut away—through spinal lesions—then Sherrington's conception of a reflexive field as a system of responses makes sense. But at the level of the whole organism and its relationship to a complex environment of multiple (and possibly conflicting) stimuli, then, multiple and variable parallelism of responses might indicate the character of a stance, a disposition, or a set of "affordances" (in J. J. Gibson's famous terminology) toward the environment by the organism. Or this might be interpretable in the more biosemiotic terms of Jakob von Uexküll's *Umwelt* (environment) ([1932] 1992), the way an organism responds to environmental cues as a sensory sign-system. The various phenomena in different fields, and according to different neural pathways and reflexes, forms a unitary whole for the organism, and are to be distanced from Sherrington's formulation of a system of reflexes: "The reactions exceed the limits set by the theory of reflexes" declares Goldstein ([1934] 2000, 176). For Goldstein, then, the argument focuses not on abstract stimuli-response mechanisms, but on the legibility or interpretation of those stimuli through sensory organs and subsystems, so that "the stimulus effect is determined by the 'functional significance' of the stimulus for an organism" (177). As we saw even in Sherrington, most of the time "nociform stimuli," and in fact any stimuli that relate to the organism as a whole, outweigh other stimuli. But there are exceptions. We noted that the flexor reflex can be suppressed if a subject requires information about an object in its environment:

> There are situations in which an individual endures pain, for example, for the sake of "higher" interest. . . . This proves that stimuli are dominant not because they are nociform, not because there may be special noci-receptive organs (Sherrington), but because this injurious effect under certain circumstances becomes more important for the organism than all other stimuli to, or actions of, the organism. Again we see how important a factor the *functional significance* of the stimulus is. (177, emphasis added)

Recent rearticulations of the idea of an organismic perspective on pain have been offered by neurophysiologist A. D. "Bud" Craig, previously invoked because of his work on "interoception," and also Malika Auvray, Erik Mylin, and Charles Spence, researchers in the psychology of the senses. First, Craig and homeostasis. Building upon ideas from, on the one hand, Charles Darwin's *The Expression of the Emotions in Man and Animals* (1872) and, on the other hand, the aforementioned Darwinian near-contemporary Claude Bernard, the concept of homeostasis as a dynamic process that involves many integrated systems is popularized by Walter Cannon in *The Wisdom of the Body* (1939). The continuous maintenance of optimal balance in the physiological condition of the body should find resonance given Sherrington and Goldstein's understanding of multiple and variable sensory signal processing during reflexes. Short, sharp pain, and pain of longer duration, such as the sensation of burning, are important signals for a homeostatic system that relies on feedback to regulate the condition of its *milieu intérieur* against alterations or differences, and to mediate the body's *milieu intérieur* with its *milieu extérieur*. As such, "changes in the mechanical, thermal, and chemical status of the tissues of the body—stimuli that can cause pain—are important first of all for the homeostatic maintenance of the body" (Craig 2003, 303). A classic illustration of homeostasis is thermoregulation in animals and humans. According to Craig, pain works not only analogously, in terms of its homeostatic mechanism, but also through common neural pathways. Consequently, in the same way that animals thermoregulate, all vertebrates respond to what Sherrington (1906) had described as the "noxious" stimuli detected by nociceptors that produce feelings of pain in humans and vertebrate nonhumans alike. Nociceptors respond only to tissue damage due to a straightforward mechanism of alteration or change, but even Sherrington acknowledged that the assessment of harm or potential for harm is not down to nociceptor reflexes alone, and so response "to a wide range of stimuli of *different* kinds was required" (1906, 228, emphasis in original). At least for more sophisticated vertebrates, as we saw, pain was not a simple sensation based on a series of reflexes but part of a more variable, and often consciously directed, series of reactions that have the potential to culminate in the movement of the organism, as Sherrington detailed with flexor suppression. In which case, nociceptor feedback is placed within a larger perceptual context of the organism's awareness of its *milieu extérieur* along with other sensory content (temperature, proprioception, fatigue, etc.), and might be considered as one among a series of homeostatic processes even within Sherrington's framework.

Since nineteenth-century approaches to pain had assumed it was physiologically inextricable from touch, Craig argues that experimental data from cats and monkeys have been conclusive in revealing "specialized central substrates that represent pain, temperature, itch, muscle ache, sensual touch and other bodily feelings as discrete sensations within a common pathway," and in fact this is a "primordial homeostatic afferent pathway that represents painful stimuli in distinct sensory channels alongside all other aspects of the condition of the body" (2003, 304). The two types of pain, what might be characterized as "fast" (urgent, sharp) and "slow" (dull or burning), may share a common afferent pathway, meaning the route of the nerves from periphery (body) to the central nervous system (in this case the spinothalamic tract, a part of the spinal cord that connects to the thalamus). As we saw in the previous section, they are conducted via different nerve types, the narrower and slower C-fibers for dull pain, and the larger, more conductive Aδ-fibers for sharp pain. But as Craig and also Auvray et al. argue, identifying the perceptual pathways of pain does not tell the whole story for the organism. Helpfully, Auvray et al. distinguish between the "sensory-discriminative" aspect of pain, which is associated with one set of brain areas, including the somatosensory thalamus and primary and secondary somatosensory cortices, whereas the "affective-motivational"

aspects of pain occur in areas usually associated with affective processing, the amygdala, the medial thalamus, and the limbic cortex (2010,
220). Referring to Craig's work on interoception, they similarly recon-
sider the import of the affective-motivational pathway at the organismic level: "Pain is also essentially *motivational* in nature" and so "provides a conscious manifestation of a pre-conscious perception of threat
to body tissues that motivates us to get our tissues out of danger" (2010,
219, emphasis in original).

    From Craig and Auvray et al., via Sherrington and Goldstein then,
pain involves a set of signals that build up an intra-organismic sequence
of responses that have the potential to tip the whole organism from
perception to action, from sensation to motility. The prompt is usually
from the outside, but the responses involve sets of autonomic responses
triggered to maintain the stasis of the *milieu intérieur*. As Craig puts it:
"Homeostatic afferents generate both a sensation and an affective motivation with autonomic sequelae—that is, a feeling from the body that
motivates behavior" (304). In organisms that thermoregulate effectively,
or experience pain, then, the subject's perception of their internal body
state (interoception) is crucial in order to return to optimal conditions.
For example, the thermal cutaneous stimulus of "warmth" or "cold"
is felt by the subject only relative to the optimal thermal conditions
of their body, usually around 37°C. A change in exterior temperature,
registered as an affect, a novel thermal cutaneous stimulus, spills over
into behavior as the subject searches for a scarf or jumper, or moves to
a warmer environment. Increasing discomfort is experienced at temperatures below 24°C, but around 15°C this is no longer registered as
temperature but as pain. From the physiological evidence, Craig argues,
homeostasis "rather than the heuristic oversimplification 'nociception,'
is the fundamental role of the small-diameter afferent fiber [i.e., C-fiber]
and lamina I system[,] and is the essential nature of pain" (306). Viewed
as part of a larger, enmeshed homeostatic system, pain can be seen as
interacting with other homeostatic functions as well as affective states,
with implications for revisiting some well-known experiential mind–
body interactions that remain philosophically problematic, the kinds
of "dependencies" that Fechner had invoked, thinks Craig, or psychosomatic illnesses where there is potential for chronic pain management
through clinical therapies (307). Moreover, the homeostatic view of pain
allows for the kinds of fast (sharp) pain, and slow (burning) pain chan-

nels, the former of which can elicit fight-or-flight behaviors, and the latter of which engages in long-term responses, sickness behavior, and immune function, each of which involves different attendant affects:

> These findings indicate that in primates a forebrain system has evolved from the hierarchical homeostatic system, and that this provides a discrete cortical image of the afferent representation of the physiological condition of the body (which we term interoception), along with direct activation of limbic motor cortex. The interoceptive system is distinct from the exteroceptive system associated with touch and movement, although there is overlap (in area 3a of the sensorimotor cortex) with respect to pain. (304)

Seen in these terms, the role of consciousness is secondary to the detection of pain and other sensations, and the motility of the organism as a whole. But the preplanning of movement, and the appraisal of the organism's environment and the range of possible actions, is certainly helped by this putative "cortical image" that need not be representational of the body or bodily position in space; instead, it might be considered allied with Head and Holmes's concept of the "body schema." Pain perturbs this schema, alters the *milieu intérieur,* but activates not only nerve pathways dedicated to nociception, as we saw, but also finds representation, prompts activity, elsewhere—including the sensorimotor cortex—eliciting in the whole organism movement, or movement-potential.

Pain and Sensation: Concluding Thoughts

Pain sensations that arise in an organism might be muscular, co-arising with movement or a shift in posture. Or they might be due to an itch, a change in temperature, arriving suddenly as a noxious stimuli as an intrusion into the exterior, or involve a rhythmic temporal cycle (throbbing), for example. When distracted, our perception of any differences in our *milieu intérieur* is lessened or impaired. Nevertheless, pain has an intensity of its own, something we experience differently from other, even related, sensations. There are specific but diffuse nervous pathways (dull or sharp, C-fiber or Aδ-fiber) spread throughout the body, and specific receptors (nociceptors) which as we saw feed into interoception. Intense pain states might prompt the organism to take action (via affective-motivational information) to urgently adjust to its external environment. While Weber, Fechner, Sherrington, Goldstein, and others in this chapter so far have measured, mapped, or considered the function of pain in various ways (e.g., as unit of measure, as reflex unit, as organism, as biosemiotic system), all of this has assumed the presence rather than absence of pain and touch stimuli. Even in cases of vague somatosensation, where sensation is difficult to distinguish against the background of what Weber termed "common sensibility," or Craig's "interoception," pain as a sensation has been induced, inflicted, measured, mapped by professionals, doctors, experts. Its presence has been prompted, in other words, from an outside.

This chapter started with Wilson et al.'s study of boredom and the introduction of autoproduced pain. Scarry has argued that, in the absence of stimuli, pain contracts the body down to a small space (1985, 35). Honkasalo observes this in her ethnography of visitors to a pain clinic in Finland, quoting her research subjects: "Space can be described as shifting and fluctuating, as 'shrinking, constricting,' but as also expanding. . . . 'The space is folding away from my hurting body'" (1998, 36). As an organism, the motivation—the stimulation from outside that triggers a series of motoric responses—is to whither, to engage in self-preserving contraction from an outside threat. This makes sense in terms of both Sherrington's mechanistic neurology and Goldstein's organismic physiology. Yet the self-administered pain that results in Wilson et al.'s study from a complete absence of external stimuli prompts us to consider the opposite perspective. Autoproduction of pain may occur not because the burden of "just thinking" weighs too heavily on a distracted or unskilled consciousness, whereby any strong sensory stimulus is preferable in disrupting the tyranny of a contracted, or relatively impoverished, sensory state. But rather, we might consider it as an organismic response to an imposition from the outside toward sensory stasis. In other words, an organismic habituation that reduces neural noise to a background bodily state. Rather than shrinking from pain, is the autostimulation of pain as sensation a way of making the organism matter to itself, produce a series of intensities, offer up a sensory threshold to consciousness, such that the organism reassures itself of its bodily boundaries, is called to take a stance toward the external world, and is subsequently made to feel "alive"? Various concepts raised so far, including Bernard's concept of *milieu intérieur,* echo in Scarry's discussion of the "projection" of sensations into determinable locations as part of human consciousness, where sight and smell become coordinated with those inanimate externalities perceived through vision and olfaction. With this projection comes a heightened awareness of some bodily core, she argues, the ability "to move further in toward the interior of felt-experience"; it is somehow the wellspring of an "awareness of aliveness" that occurs at some level within "the felt-experience of sentience" (1985, 285). In this schema, is self-administered pain not just the production of sensation within a determinable location on the surface of a body, but somehow—or, simultaneously—a stimulation routed through the bodily interior that reaffirms the sentience of the body against its overall tendency toward homeostatic and sensory habituation?

Finally, another related pathway opens up for understanding Wilson et al.'s experimental subjects. The physiologists and psychophysicists encountered so far assume that organisms are pain-averse, and consequently that a series of responses and reflexes are adverse reactions, motoric pathways, and conscious means to restore an internal milieu. Yet we also acknowledge a long-running rationale for the positive aspects, and even enjoyability, of pain. Something allied to, but never completely reducible to, masochism. For example, the first English dissertation on the subject, *An Essay on the Nature of Pain* (1826) by a member of the Royal College of Surgeons in Edinburgh, William Griffin (1822), acknowledges this aspect. As does a chapter "The Enjoyment of Pain" by the Finnish scholar Yrjö Hirn in his *Origins of Art* (1900) that seems uncannily to anticipate Wilson's study:

> If we take into account the powerful stimulating effect which is produced by acute pain, we may easily understand why people submit to momentary unpleasantness for the sake of enjoying the subsequent excitement. . . . The creation of pain-sensations may be explained as a desperate device for enhancing the intensity of the emotional state. (Hirn, quoted in Mososco 2012, 84)

Xavier Bichat had also argued about the perceiving subject's adaptability to pain as, like other sensations, it involved a comparison between different states, but also that they tended to disappear over time. Bodies in pain therefore have a tendency to incrementally retreat into a normative background bodily state as part of the process of habituation. In a section entitled "Habit Blunts the Sentiment" of his *Physiological Researches on Life and Death* (1800), for example, Bichat prefigures Maine de Biran's paradoxical formulation of habit in stating: "Pain and pleasure are absolute and relative" ([1800] 1815, 53). These sensations are experienced as absolute at the time, yet are relative once enough time has passed for the subject to become habituated. Bichat illustrates the arc of sharp pain and subsequent overall organismic habituation in rather eye-popping style: "A bougie [tapered probe] when for the first time passed into the urethra is painful to the patient; eight days afterwards he is no longer sensible of it" (53–54). Given the incredible variability of the subjective experience of pain, and the fact that pain impairs judgments as to the origin and even the location of its original source, the days of introspection as an experimental methodology were always numbered. Inevitably, like other sensory phenomena, the articulation and legibility of signs of pain would shift to other forms of graphical inscription. In medicine and neurology, neuroanatomical explanations through fMRI scans would bypass mere subjective reporting of interoceptive pain states. Of course, fibromyalgia and chronic pain states persist, despite the scans. Intensities without proximate causes. However, the question posed at the start of this chapter, and addressed through histories of experimentation and physiological theories of the organism, concerns no immediately visible or invisible lesion, nor any potential neurological dysfunction. It has involved the consideration of pain as an autoproduction of sensation against a background of overall organismic habituation, at least for neuroanatomically normate human subjects, and for a variety of motivations.

The kinds of experimentation and self-experimentation we have discussed may therefore be far from the simple aversion to the boredom or "default mode" of "just thinking" that Wilson's team proposed. As a means of managing our own affective intensities, and in underlining the "social" within the biopsychosocial, the relationship between pain and sensation exceeds the boundaries of the individual body.

CHAPTER 3

The Oculomotor

Labyrinths, Vestibules, and Chambers

How is the inner ear involved in sensations of movement and balance? And what has this to do with the involuntary rapid movements of the eyeball? The clusters of organs and nervous pathways that make up the vestibular system were pursued independently by two scientists from the 1870s onward, one in Vienna and the other in Prague, and eventually unite through what became known as the Mach–Breuer theory of semicircular canal function. Until that point, the gross anatomy of the vestibule, semicircular canals, and cochlea of the osseous labyrinth of the internal ear had been known since Galen in the second century, and its internal fluid (endolymph) investigated in the seventeenth century by the Italian scientists Antonio Valsalva, Domenico Contugno, and Antonio Scarpa, and in the early nineteenth century by Alfonso Corti. But the function of this inner osseous architecture was erroneously and vaguely identified exclusively with auditory capacities until experiments by the physicist Ernst Mach in Prague in the 1870s. Mach investigated the nonauditory function of this spatially entwined organ cluster, demonstrating both that the vestibular (balance) system of the inner ear was involved in sensations of movement, and that eye movement affected motion sickness. But it was the physician and physiologist Josef Breuer's work on the sensory receptors of the inner ear, in a series of papers between 1873 and 1903, that definitively proved its nonauditory function. In 1873 Breuer delivered a paper to the Imperial Society of Physicians in Vienna, "On the Function of the Semicircular Canals of the Aural Labyrinth," identifying the role of the canals in sensing angular movements of the head as a result of the movement of endolymph within the three planar orientations of the canals. Further, Breuer identified a crucial mechanism tying the stimulation of the semicircular canals by movement to automatic compensations by the eyes, such that when the body spins in one direction the eyes are immediately directed toward the opposite direction. This phenomenon is well under-stood in recent neuroscientific literature as the vestibular-ocular reflex (VOR). At the same time that Breuer was conducting his experiments in Vienna, Mach was working on a series of his own experiments in Prague with a spinning chair, published as *Grundlinien der Lehre von den Bewegungsempfindungen (Fundamentals of Movement Perception)* in 1875. Breuer and Mach corresponded over their discoveries, leading to what became known as the Mach–Breuer theory of semicircular canal function. The significance of Breuer's and Mach's work continued into the era of early astronaut training in the mid-twentieth century, and the vestibular-ocular reflex's (VOR) role in motion sickness, as we shall see.

While this work on the developing neurophysiology of the inner ear in the laboratories of Mach, Breuer, and Wundt in the nineteenth century was foundational, I also present contemporaneous developments in the aesthetic experience of the pedestrian, or *moving,* subject. The motility of the subject is often downplayed in the appraisal of aesthetic objects and built environments. Yet, around this time, writing in art history and the history of architecture began attending to the oculomotor element within perception. The distinctive spiral-spatial forms of these nested organs within the bodily interior, given their analogic associations with the space of labyrinths, vestibules, and chambers, find their real-world counterparts in architectural spaces outside of the body, involving a heightened physicality and sensory engagement with the built environ-ment. This mapping between the chambers and spaces of the inner ear to those of spaces and environments external to the body starts with Mach, Breuer, and Wundt, but will involve near contemporaneous pioneering work by the Swiss Heinrich Wölfflin, the Germans Wilhelm Worringer and Johannes Volkelt, and others such as the Austrian Aloïs Riegl, that incorporated findings in psychology to complicate the realm of aesthetic experience beyond mere vision or static ocularity. Riegl ([1901] 1995) famously distinguished between the "haptic" and "optic" in art history but, when it comes to historicized accounts of spatial perception, as Martin Jay (1994) and Jonathan Crary (1992) show, the complexities of sensation remained for the most part obscured by the continued dominance of a static model of vision as the principal orga-nizing modality. As we shall see, Erwin Panofsky addresses this in part in his *Perspective as Symbolic Form* (1913), where the celebrated pioneer-ing of the static form of homogenized linear perspective attributed to figures such as Brunelleschi is, he says, "quite unlike the structure of psychophysiological space" ([1927] 1991, 29), and "forgets that we see not with a single fixed eye but with two constantly moving eyes" (31). It just so happened that, not long before Panofsky was writing, an archaeological dig in North Africa was conducted by Alfred Merlin (1917), and features of a Roman complex of gymnasium and baths for the training of wrestlers known as a palaestra were being pieced together. This environment would offer visitors an experience of heightened physicality not only through the athletic activities on offer but also through the use of floor mosaics and features underfoot that also drew in the eye and, through the complex of pools at different temperatures, offered thermic pleasures for the skin. This chapter therefore returns to prior historical articulations of the significance of motility in perception to develop an approach which enriches visuality beyond static ocularity, through the entwined neurophysiological mechanisms that involve the vestibular system, the motility of the embodied observer, and the saccades of the eyes. This is a realm of aesthetic experience that I have characterized elsewhere as "more-than visual" (Paterson 2011).

To this end, the chapter is composed of four sections. First, I summarize some recent tensions between haptic and optic, and what has been termed "retinal vision," initially within architectural writing. By considering some of the physiological contributions involved in perceiving certain historical artworks and architecture, the functions of oculomotor perception are opposed to the merely "retinal" model of vision. The shift from the putative static model of the eye to an investigation of more motile ocular processes, as I argue in the second section, "Haptic Perception," is supported by some innovative early twentieth-century work bridging psychology and art history, where a particularly productive period of interchange accommodated ideas about eye movements and the production of the visual field, especially for Heinrich Wölfflin. From this historical interchange, and to increase distance from more traditional models of static vision, I proffer a particular form of heightened physicality that occurs within the visual encounter, effected through a newly oculomotor subject. Third, in "Labyrinths, Vestibules, and Oculomotor Processes," I connect with more contemporary neurophysiological work on the oculomotor system to further unfold this line of inquiry, and outline research on the established neurophysiological relationship between the mechanisms of the inner ear and the movement of the eyes in three-dimensional spatial perception. Here, the deep integration between these subsystems is discussed, known in neuroscience literature as the vestibular-oculomotor reflex (VOR) (see, e.g., Lackner and DiZio 2005; Cullen 2012). Finally, in the fourth section, "Motility of Eye and Feet," these strands come together through the archaeological case study of a built environment, the mosaic spaces of the Roman palaestra. Here, writing on sculpture and the sculptural process, and then art historical investigation about the heightened physicality involved when encountering Roman floor mosaics, are brought together via the eye and the foot of the oculomotor subject. Although this type of embodied trajectory has been accomplished by appeal to phenomenology in recent years, I reinvoke the psychologically oriented art historical writing from the second section. This includes Wölfflin's "Prolegomena to a Psychology of Architecture" (1886) and his rejection of the straightforward idea, following psychologist Wilhelm Wundt, that "the emotional tone of a form is explained by the *kinaesthetic response of the eye* when its focus follows the lines" ([1886] 1993, 150, emphasis in original). As this chapter progresses, spatial motifs of chambers, vestibules, and labyrinths loosely, but opportunely, map those interior spaces of the body involved in vision, balance, and orientation onto their architectural namesakes, those features and aesthetic spaces that enhance—almost exaggerate—the physicality of the observer, an oculomotor subject.

Retinal Vision

It is now commonplace to acknowledge that models of vision within the popular scientific imagination are indebted to legacies within intellectual history such as Descartes's *Dioptrique* (1647, On optics), where the static model of the camera obscura functioned as an explanatory mechanism not only for the visual process but also as an epistemological model for a disembodied subject. This same idea would later find expression in what Locke called the "dark room" in his *Essay Concerning Human Understanding* (1690), and what Newton similarly termed "a very dark Chamber" in his *Opticks* (1704, in Crary 1992, 246). These familiar analogues affirm the centrality of visual processes and of the eye as the organ of vision, an idea memorably termed "ocularcentrism" by Martin Jay (1994). Unlike those models, which involve light hitting the back of the retina of an idealized static subject, at once disclosing the truths of proportion and perspective to an observer, in this first section I open up the investigation to a wider series of ocular and motor processes that together help constitute alternative figurations of visuality for mobile, embodied subjects. How such ocular processes involve other sensory subsystems, including the semicircular canals and otoliths of the inner ear that make up the vestibular system, will reveal mechanisms that bind balance, bodily movement, and the constantly fluctuating and renewed visual field. This line of inquiry, placing the body back into aesthetics and the perception of architectural space, has been broadly compatible with a phenomenological psychology (especially in Maurice Merleau-Ponty 1964, 1992), and the phenomenological work that has arisen within architecture by Stephen Holl, Juhanni Pallasmaa, and Alberto Péréz-Gómez (2006) and Jonathan Hale (2013), for example. Instead, given the near-contemporaneous developments in the history of psychology and the development of a nascent art history at the end of the nineteenth century and the beginning of the twentieth, here I discuss this in proto-phenomenological terms couched in the scientific language of the time, and consequently in terms of the oculomotor system, the compensatory eye movements that result from the integration between the vestibular system of the inner ear, and the movement of the eyes through direction of extraocular muscles by motor neurons.

To establish what is meant by a set of entwined oculomotor processes, I start by laying out a knowingly outdated dichotomy. On the one hand, following historical discussion of the camera obscura, a common assumption in art history and architecture is of a stale, static model of vision that might be termed "retinal" by the architects Le Corbusier and Juhanni Pallasmaa, and the "retinal image" by art historian Erwin Panofsky. On the other hand, there are historical precursors for current empirical research on the entwining of ocular processes with other sensory and motor subsystems, including balance, and as discussed in the next section, crossovers between psychology, art history, and architectural theory in the late eighteenth and early twentieth centuries. To set up the first part of the dichotomy, then, let us consider a bald but representative statement about the role of vision in architecture by one of its preeminent modernist practitioners, Le Corbusier. At one point in *Vers un architecture* (1923) (*Towards a New Architecture,* 1986) he confidently declares:

Architecture is the masterly, correct and magnificent play of volumes brought together in light. Our eyes were made for seeing forms in light; shadow and light reveal forms; cubes, cones, spheres, cylinders, and pyramids are the great primary forms that light reveals well; the image is clear and tangible for us, without ambiguity. (Le Corbusier 1986, 102)

Fascinated by the geometry of volumes and surfaces, Le Corbusier here encapsulates the particular model of vision from Descartes and Newton that might be described as "retinal," the ostensibly straightforward movement of light and shadow from the surface of an object consequently playing out on the surface of the retina. It is easy to caricature his unsophisticated model of vision, one that asserts a seemingly unambiguous nature. This model of vision is rarely challenged in everyday belief. Nevertheless, in a passage that celebrates such purely visual properties, an opposing thesis might be recovered instead about the *ambiguities* of vision, of the difficulties of separating opticality from hapticality for a mobile, viewing subject.

Yet the fascination with eyes, screens, and flat retinal images by architects persists. Recently it has been echoed from a more critical perspective, the practice and theory of architecture being described in explicitly visual terms as "an interlocking of frames" and "a cinema of things" by Bernard Cache (1995, 29), and more pejoratively as merely a "logic of the image" (1998, 289) or as "retinal architecture" (2005, 26) by another, Juhani Pallasmaa. Arguing directly against Le Corbusier's visuocentric perspective, for example, Pallasmaa claims that the use of various design technologies has only furthered "the detachment of construction from the realities of matter and craft," a process that turns architecture into "stage sets for the eye" (Pallasmaa 2006, 29). Hartoonian also speaks of "the marginalisation of the tectonic and tactile dimensions of construction" (2001, 54) since Le Corbusier. So far, this would seem to suggest rather convenient tensions between an assumed ocularcentrism and the types of nonvisual sensations and somatic practices that have been hitherto ignored or underexplored in architectural thinking and design practice alike. But, as the following section will help problematize, the "haptic" has increasingly begun to stand for something nonspecific that simply opposes the centrality of visual experience within art, film, and architectural theory. Against what he identifies as the "dominant retinal architecture," for example, Pallasmaa celebrates an emerging "haptic architecture" of figures like fellow Finnish architect Alvar Aalto. In an essay marking the significance of Aalto's contribution to architectural practice, for example, Pallasmaa speaks of the "hegemony of the eye over the other sensory realms" and bemoans the fact that "architecture has turned into the artform of an instant visual image" (1998, 296). But due to its purposive contact with the skin, an object's tactile qualities should also be considered in the design process, he argues, even in terms of the temperature and thermal conductivity of the material. For Pallasmaa, at least, an architectural "haptic sensibility" therefore emerges:

> The haptic sensibility suppresses the dominance of the visual image through enhancing plasticity, tactility and intimacy. An unconscious element of touch is unavoidably concealed in the sense of vision; as we look so the eye touches the object. This hidden tactile experience determines the sensual quality of the object, and mediates messages of invitation or rejection, home or hostility. (297)

Notably, his "haptic sensibility" is not simply set up in opposition to "retinal architecture," as visual experience is expanded beyond the merely optic. Furthermore, while the relationship between "haptic" and "optic" is briefly discussed in the next section, I will not be making great claims for the power of the haptic within the experience of art, architecture, or film, an idea that was previously in vogue (see, e.g., Paterson 2017 for a critique of so-called haptic visuality). Instead, my purpose here is to uncover contemporaneous explanations within art history and psychology for how oculomotor perception encompasses a range of ocular processes that are neither strictly "haptic" nor "optic," but that are instead a set of more-than-visual perceptions of a motile body. This genealogy unfolds through the ages as a series of sensory–spatial motifs, effectively an associative mapping between the chambers and spaces of the inner ear to those of spaces and environments external to the body.

In its original historical context, the concentration of the interplay of light from surfaces, interpreted by Le Corbusier as producing unambiguous, distinct, optically delineated forms, would only seem to further the ocularcentric approach to the design of buildings. For the play of light that evoked surfaces, volumes, and depths is straightforwardly demonstrated through the cuboid apparatus of the camera obscura. Furthermore, his description of unambiguous vision does equate the projection of light into the eye with that of a static canvas. As Aloïs Riegl claims, "Every work of art does presuppose the existence of a perceiving subject" ([1902] 1988, 181). In the case of painting, epitomized by Leon Battista Alberti's formalized conventions of Western perspective, a single, immobile point near the center of the work is the location the perceiving subject must assume in order to experience the artist's intended spatial effects. Conceiving of the dimensionality of material forms in this commonly understood way, through Alberti and Le Corbusier, is a simple matter of painting with light in a static, monocular fashion that might be designated "retinal." As with the camera obscura, the implication is that, through some unnamed mechanism or unrecognized perceptual process, we intuitively translate two-dimensional surface into three-dimensional shape. Yet, unbeknownst to him, Le Corbusier's language allows for a more ambiguous kind of visual model than he intends, especially regarding how the image becomes, in his words, "tangible for us" within the visual field. Transposed from surfaces, canvases, or walls onto the decidedly nonplanar concave surfaces of the retina, and with surrounding extraocular muscles that rapidly orient the eyeball, would prompt us to consider that the ocular processes involved in constituting a visual field are anything but static or flat.

Only a few years after Le Corbusier's words, Panofsky's *Perspective as Symbolic Form* (1927) commenced with a similar idea, of an abstract perspective in which all perpendiculars and orthogonals meet at one point, the central vanishing point, supposedly providing a "fully rational," "infinite, unchanging and homogenous" space ([1927] 1991, 28–29). By means of a diagram of idealized linear perspective, where lines of sight converge at the single point of an imaginary observer as a perfect pyramid (see Figure 8), Panofsky soon dismisses this idealized construction and directly contests the assumptions behind retinal vision. Arguing diametrically against Le Corbusier's naïve visual model, in fact, Panofsky maintains that perspective is never apprehended through direct sensory representations, but always through bodily abstractions that are taken as erroneous assumptions: first, that "we see with a single and immobile eye," and second, that, for example, "the planar cross section of the visual pyramid can pass for an adequate reproduction of our optical image" (28). In other words, this is not the "unambiguous" and clear visual perception of a pyramid that Le Corbusier had assumed. The perfectly straight lines from the diagram, in this case illustrating a three-dimensional shape of a "space box," is simply not what we phenomenologically experience, and so "is quite unlike the structure of psychophysiological space" (30) we obtain from our senses. Whereas the mathematical space of perspective is isotropic and homogenous, quoting Ernst Cassirer's *Philosophy of Symbolic Forms,* he argues: "Visual space and [tactile] space *[Tastraum]* are both anisoptropic and unhomogenous" (30), as there is clearly a psychophysiological difference between front and back, left and right, from the point of view of the embodied observer, a point that Ernst Mach makes and Panofsky quotes briefly without citing him (see Elkins 1994, 191ff for a critical perspective on Panofsky's argument here). Panofsky could be chiding Le Corbusier directly when he writes of "perspectival construction," with its transformation of psychophysiological space into mathematical space:

> It forgets that we see not with a single fixed eye but with two constantly moving eyes, resulting in a spheroidal field of vision. It takes no account of the enormous difference between the psychologically conditioned "visual image" through which the visible world is brought to our consciousness, and the mechanically conditioned "retinal image" which paints itself upon our physical eye. (31)

This pattern of light on the retina is somewhat stabilized through the cooperation of touch, argues Panofsky, where the size and form of objects can be collocated and resolved into a recognizable form. He advances a further problem about perspectival construction, however, as it "ignores the crucial circumstance that this retinal image—entirely apart from its subsequent psychological 'interpretation,' and even apart from the fact that the eyes move—is a projection not on a flat but on a concave surface" (31). The projection of straight lines is habitually perceived by the eye as convex curves, and according to Panofsky, "the curvatures of our, so to speak, spheroidal optical world had to be rediscovered" (34), which it was by psychologists and physicists in the nineteenth century such as Helmholtz in his *Handbuch der physiologischen Optik* (1867, Handbook of physiological optics) and Wundt's *Grundriss der Psychologie* (1896, Outlines of psychology), but also mathematicians and astronomers at the beginning of the sixteenth century.

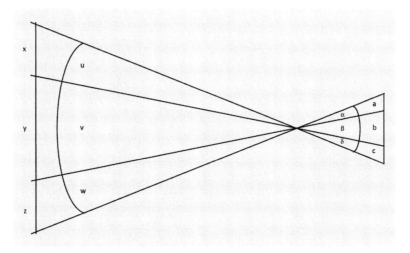

FIGURE 8. *Explanation of marginal distortions. If a line is divided so that its three sections subtend equal angles, these sections will be represented on a concave surface as approximately equal lengths; if projected on a flat surface they will appear as unequal lengths. Adapted from Panofsky (1927) 1991.*

Part of the spheroidal field of vision is a result of the curved plane of the retina, for Panofsky, but prior to what would become the psychological study of eye saccades he had mentioned the role of "two constantly moving eyes" ([1927] 1991, 31). What if the reassuringly static model of the camera obscura with its planar rear wall upon which light is received were to be reconceived according to current understandings of how the eye moves rapidly through saccades in order to produce the experienced visual field? Accordingly, what if the concave "canvas," even the entirety of the camera obscura (originally, a whole darkened room), were to move rapidly, darting between features in scenes? For in actuality the *orbis,* its movement directed by six extraocular muscles (the fastest-moving muscles in the human body), does not passively receive an "unambiguous" scene, but rapidly and continually glimpses, recomposes, remakes visual data, piecing it together in order to constitute a continually experienced "visual field" (see, e.g., Burr and Morrone 2004 on their groundbreaking work on saccades). The traditional idea of perspective therefore denies "that the essentially physiological construction of vision sets a narrow cone of foveal vision and a rapidly moving eye against a broadly peripheral [visual] field," as Marc Treib (2007, 62) summarizes in his essay on the perception of landscape. The inherently oculomotor nature of the perceiving subject becomes manifest as they physically proceed through a built environment, perceiving walls, decorations, porticos, and so on as a rapidly shifting pattern of light and shade that continually refreshes the visual field, making the shapes and volumes appear consistently and continually anew. The retinal image in Le Corbusier's terms made "tangible for us" entails a heightened physicality of the body, therefore, a progression from a static or "retinal" model of vision through the movement of the entire body, including the textures of ground underfoot. Although use of the word "retinal" need not be associated exclusively with the receptive surface of a static organ, since the eyeball is constantly moving, the suggestion of an oculomotor model of visual processes underlines the often ambulatory, and always multisensory, nature of embodied perception. In architectural settings that invite movement, as opposed to a static canvas, materials underfoot as well as optical effects are at play. This is perhaps the form of "tactility" that Taussig intends in his essay on Benjamin and architecture, "Tactility and Us," where it is habit and tactility that makes the optical experience of a building, as Le Corbusier put it, "tangible for us":

> To the question "How in our everyday lives do we know or perceive a building?" Benjamin answers through usage, meaning, to some crucial extent, through touch, or better still, we might want to say, by *proprioception,* and this to the degree that this tactility, constituting habit, exerts a decisive impact on optical reception. (2007, 261, emphasis added)

A rewired tactile optics, based on habitual knowledge, is posed by Taussig via the Benjaminian optical unconscious. But the endpoint, what he later phrases as "tactile eye and ocular grasp" (265), is something achieved here through other means, other organs. One way to illustrate this perspective is through the example of the palaestra, a Roman wrestling school within gymnasium complexes, which typically had floor mosaics depicting labyrinths within courtyards. This example has the virtue of offering, in the words of art historian Rebecca Molholt, "representations of spatial experience that unify art and architecture" (2011, 288). After Taussig and Benjamin, then, in exploring built environments for proprioceptive subjects we again depart from any purported standard, static model of a camera obscura. Instead, in the third section I proffer an anatomical analogue with the labyrinth of the inner ear, a subsystem that, it will turn out, has a hugely significant role in visual perception within the moving body, the vestibular system subserving the oculomotor system in the movement of head and eyes.

As we know, other nonvisual sensations, such as kinaesthesia (the sense of movement), and proprioception (the sense of bodily position in space), are similarly at play for ambulatory subjects (see chapter 1). But, as is explored in more detail below, their status as distinct sense modalities is questionable due to their integrated role within sensory subsystems that deal with movement, balance, and orientation feedback. For example, after a brief survey of the five senses familiar to us from the days of Aristotle, and leading directly on from a discussion of touch, Richard Shusterman distinguishes between the two related somatic senses of proprioception and kinaesthesia:

> Proprioception concerns the inner sensations and resulting cognition of the position, posture, weight, orientation, balance, and internal pressures of one's body and bodily members, while kinaesthetic perception more specifically relates to such internally perceived feelings and resulting cognitions that relate to postural, orientational, pressure, and equilibrium changes through movement. (2011, 154)

The production of perceptions of balance and orientation is, however, the function of another sensory subsystem, the vestibular system. The lesser-discussed organs of the inner ear, components of which include a "labyrinth" and a "vestibule," are directly involved in the body's perception of its own internal states, what Shusterman has termed "somaesthesis," Heller-Roazen "inner touch" (2007), and Craig (2002, after Sherrington 1906) "interoception."

Haptic Perception

In *The Architecture of Humanism* (1914), the architectural historian Geoffrey Scott made the broad claim that because weight, pressure, and resistance are part of everyday embodied experience, these qualities were inevitably projected onto architectural forms. However, before either Scott or Panofsky, more fine-grained discoveries at the interface between psychology and art history regarding the sensory component of bodily experience, and especially the relationship between touch and vision, had already been a concern for the influential art historians Heinrich Wölfflin (1864–1945), Wilhelm Worringer (1881–1965), and Bernard Berenson (1865–1959). Although Candlin (2010) shows that Riegl, Berenson, and Wölfflin each understood to some extent the ambiguities and interchanges involved in the modalities of vision and touch, this section concentrates on the ways these figures look to physiological mechanisms that underpin aesthetic encounters, especially with architectural forms, without reducing the significance of their felt, bodily qualities. As Vlad Ionescu argues, even their precursors in architectural theory—namely, Theodor Lipps (1851–1914), Friedrich Theodor Vischer (1807–87), and Johannes Volkelt (1848–1930)—were already relating visual schemes to the affective experience of the body and attempting to escape the usual trope of "bracketing the visual as an autonomous sphere that justifies the modern aesthetic regime" (Ionescu 2016, 1). What this movement beyond the purely visual modality might mean in terms of artistic and architectural experience is the focus of this section. But first, some prefatory remarks on the physiology of bodily sensation, and what is meant specifically by "haptic."

As was discussed in chapter 1, Charles Bell in 1826 had first identified something he imprecisely termed the "muscle sense," whose character as afferent feedback (i.e., reporting from inside the body back to the brain) was later identified by Charles Henry Bastian as "kinaesthesia" in 1880. In 1906 the world-famous neurologist Charles Sherrington acknowledged "proprioception" as "the sense by which the body knows itself, judges with perfect, automatic and instantaneous precision the position and motion of all its movable parts, their relations to one another, and their alignment in space" (in Gutman 2007, 176). What some cognitive scientists currently call the "somatosensory system" is another way of formulating this set of internally felt bodily sensations. As Dijkerman and de Haan explain, the somatosensory system "provides information about the position of different parts of the body with respect to one another. It allows characterisation and localisation of touch, stroking, and pain, and it is important for all motor action involving the body and limbs" (2007, 189).

Prior to this articulation in terms of somatosensation, the distribution of bodily sensations was categorized in terms of their inward/outward orientations. For example, Sherrington (1906) considered the senses of taste and smell as "chemoreceptive," the sense of limb position as "proprioceptive," but touch, including pain and temperature, was "exteroceptive." What was considered "interoceptive" for Sherrington were visceral sensations, although this term was later widened to the physiological condition of the entire body, not just the viscera (Craig 2002, 655). A comparable orientational distinction of sensory subsystems was later taken up by ecological psychologist J. J. Gibson (1966, 1979), but Gibson considered the perception of bodily sensations including touch and proprioception as part of a larger "haptic system":

> The haptic system . . . is an apparatus by which the individual gets information about both his environment and his body. He feels an object relative to the body and the body relative to an object. It is the perceptual system by which animals and men are *literally* in touch with the environment. (Gibson 1968, 98, emphasis in original)

For the remainder of this section, I wish to highlight some important non-optical functions, the entwining of the haptic with ocular processes. For this, some familiar territory in art history is briefly revisited, albeit as a step toward identifying an art historical precursor to oculomotor activity.

Aloïs Riegl's famous distinction between the haptic and the optic (1901), so memorably revisited by Deleuze and Guattari in *A Thousand Plateaus* (1988), is pivotal. Near contemporaneously, Bernhard Berenson's argument, that when the eyes of the viewing subject alight upon an object or shape they are "giving tactile values to retinal impressions" (1906, 4), seems already to hint at the more-than-visual nature of ocular processes. However, much written during this period in art history about touch and its relation to vision is frustratingly nonspecific about the modalities of touch and vision in terms of either the character of the experience, or their physiology or psychology. Neither does it treat these modalities in their specific cultural or even historical contexts. For example, Candlin (2006, 2010) has unveiled the somewhat dubious assumptions about touch and its relationship to vision of Riegl, Berenson, and Panofsky. For Riegl, the historical precedence of tactile as opposed to optic appreciation is never assigned to specific periods, and there is no account of how or why such perceptual change supposedly occurred. In some cases, touch becomes effectively a subset of vision for Riegl, whereas elsewhere "touch is not colonizing vision but being co-opted by sight," says`Candlin (2006, 140). For Berenson in his *Florentine Painters of the Renaissance* of 1896, much emphasis is placed on illusions of touch and appeals to a putative "tactile imagination," whereas his later book of 1948, *Aesthetics and Art History,* strays from tactility to invoke more dispersed bodily sensations, reminding us of Scott's earlier importation of the bodily into the architectural: "In order to be life-enhancing an object must appeal to the whole of one's being, to one's senses, nerves, muscles, viscera, and to one's feeling for direction, for support and weight, for balance, for stresses and counter stresses," says Berenson (1948, 58). Sherrington's orientational categorization of the senses, the interoceptive and the exteroceptive, is also implicated in the way that the range of internal sensations are appealed to here.

But it is Heinrich Wölfflin who, for the purposes of this chapter, most successfully expands the interests of art scholarship beyond the retinal surface of the canvas to saccadic, and therefore oculomotor, processes. Wölfflin's precursors were engaged with the concept of empathy *(Einfühlung)* in art history, a central idea for Theodore Lipps's *Ästhetischen Faktoren der Raumanschauung* (1891, Aesthetic factors of space-intuition). Empathy is not simply the projection of one's feelings onto an object but a participative emotional immersion into an observed object, the pleasure of a subject who "feels his/her way into" an object. In German the term *hineinfühlen* infers a forward movement of that observing subject. Reminding us of the emergence of a distinctly muscular sensation in chapter 1, Johannes Volkelt in *Der Symbolbegriff in der neuesten Ästhetik* (1876) (1876, The concept of the symbol in the latest aesthetics) writes of the connection between visual experience and an embodied sense of vitality, a human vital feeling *(Lebensgefühl).* He argued, for example, that our "bodily organisation takes part into *[mitmachen]* the experience of spatial constructions that are sensuously apprehended *[miterleben]*" (Volkelt 1876, 57). This idea of the bodily organization of spatial experience will be revisited by Wölfflin later in this section, but it also prefigures an extensive discussion of the spatialities of one's own body *(le corps propre)* in Maurice Merleau-Ponty, as I show in chapter 6.

Although Wölfflin is known primarily as an art historian with classic texts including *Renaissance und Barock* ([1888] 1964, translated as *Renaissance and Baroque*), his contributions to psychology and architecture were cemented in his first publication, his doctoral thesis,

Prolegomena zu einer Psychologie der Architektur ([1886] 1993, translated as "Prolegomena to a Psychology of Architecture"). He opens the
Prolegomena with a straightforward albeit pointed question: How is it
possible that an inanimate object (a building) conveys an impression
(Eindrück) that is felt as an expression *(Ausdrück)*? Somewhat mirroring
our discussion in the first section, Wölfflin debunks some of the assumptions behind "retinal vision." Along with the work of Lotze, Wölfflin specifically targets Wundt's physiological explanation, according to which
the impression of visual forms depends on the movement of the eye. A
little background to Wölfflin's remarks on Wundt are helpful here.

Wundt had argued in his *Vorlesungen über die Menschen- und Thier-
Seele* ([1863] 1896, translated as *Lectures on Human and Animal Psychology*), in a lecture entitled "Influence of Ocular Movement on Spatial
Vision," that "movement-sensations" of the eye were crucial for the recognition of external objects, especially as the retinal image is in fact
originally inverted as well as curved. There must be some capacity
for the mind to "see" or translate the image into the orientation that
we ordinarily experience. Accompanying a diagram rather similar to
Panofsky's, and that clearly shows the concave nature of the retinal
image, Wundt explains:

> Only by the movement-sensations of the eye does the mind learn
> to connect these with a spatial order. But what do the movement-
> sensations tell us about the position of objects? As the eye moves it
> passes from point to point of an external object. . . . It brings all the
> parts of its retinal image successively upon the spot of clearest vi-
> sion. ([1863] 1896, 164)

That spot of clearest vision is, as we now know, the part of the retina
known as the macula. In one of his most renowned texts, *Grundriss der
Psychologie* ([1896] 1902, translated as *Outlines of Psychology*), Wundt revisits and develops the idea of ocular movement in a section on "Spacial
[*sic*] Sight Ideas," and posits a now-familiar relationship between touch
and sight whereby the fine tactile discrimination of the fingertip is analo-
gized with the "middle of the field of vision" (130), or in modern parlance, the macula. But he goes further than merely noting the analogy
between sensory organs, since the movement of both eye and hands

are similarly constrained by effort and the limitations of the musculo-skeletal frame. For one thing, he acknowledges the continuous process whereby eye movements ensure the retinal image is located at the concave center of the back of the eye: "The location at some distance or other, which also is never absent, results in the fact that all the points of the field of vision seem to be arranged on the *surface of a concave hemisphere* the centre of which is the point of orientation, or, in monocular vision, the centre of the eye's rotation" (151, emphasis in original). For another, in both texts Wundt recognizes a form of muscular effort, or even a highly localized muscle sense, at work in the muscles surrounding the eye, which bears on certain movements over others. In the earlier *Vorlesungen,* Wundt claims it is easier for the moving eye to trace the form of an object, which Wölfflin latches on to: "Owing to its physiological structure, when the eye moves freely, it follows a straight line in the vertical and horizontal directions, but it travels in an arc when moving in any oblique direction" (Wundt 1863, 2, 80; in Wölfflin [1886] 1993, 150). Likewise, in the *Grundriss,* he states: "Movements in the direction which correspond to the position of the objects most frequently and closely observed, namely, movements downward and inward, are favored above the others by the arrangement of the muscles" ([1896] 1902, 133). In this section of the *Grundriss* on spatial ideas, the perception of distant objects involves the movement of the eyeballs in a coordinated manner, as a result of the musculoskeletal arrangements of the extraocular muscles, including the *superior rectis* and opposing *inferior rectis,* which both help rotate the eyeball in the horizontal plane, activated through the oculomotor nerve. If Wundt argues that certain eye movements are favored over others, it is because of a difference in motor energy, and a concomitant sense of strain in the perceiver. Visual space and the perception of objects at various distances for Wundt therefore also incurs subjectively felt artifacts within ocular processes, including "the intensive motor sensations determined by the relation of the stimulated points to the centre of the retina" (143). Similarly, when a perceiving subject correctly aligns his or her eyes with an object in space, obtaining "an idea of the direction of the line of orientation," Wundt claims this is somehow sensed by the subject, that this "idea is produced by the inner tactual sensations arising from the position of the two eyes" (144).

Wölfflin refers to Wundt's *Vorselungen,* not the *Grundriss,* but he is careful not to uncritically infer aesthetic criteria from such physiological observations from the 1863 lectures. From Wundt one could take the erroneous position, for example, that a picture with a regularly undulating line is more pleasing to the eye than the jagged lines of a zigzag, because in the latter case ocular movements involve more strenuous muscular motion. Were that to be the case one could make generalizing and incorrect statements such as: "The beauty of a form is directly proportional to its suitability for our eye," warns Wölfflin ([1886] 1993, 150). In fact, Wölfflin argues that this idea from Wundt is reductive, that it fails to conform with our experience. "How much of the form's actual impression can be explained by the kinaesthetic response?" he asks. "Is the greater or lesser ease with which the eye performs its movement to be regarded as the crucial factor in a multitude of effects?" (150). This clearly does not conform with our experience. His rhetorical questioning powerfully resists a tendency to reduce aesthetic inquiry to the physiological constitution of the body. Yet Wölfflin's line of questioning remains relevant if, as Ionescu suggests, "one considers contemporary research projects of neuro-aesthetics that also tend to explain beauty in terms of the neurological structure of the brain" (Ionescu 2016, 3).

In a later text of 1915, "The Linear and the Painterly," Wölfflin offers another argument that verges on a distinct explanation of the role of oculomotor processes. It offers resources for thinking about texture, as well as the planes and volumes that would soon interest Le Corbusier. This essay centers on an opposition between two modes of visuality, on the one hand the "linear" as optic and saccadic, and on the other the "painterly" as nonoptic and equivalent to the more-than-visual. While writing mostly on painting as "the art of representation on flat surfaces," his notion of the "painterly" also applies equally to different art forms, from the two-dimensional canvas to three-dimensional sculpture and architecture, so "means as much for architecture as for the arts of the imitation of nature [i.e., painting and sculpture]" (2003, 56). In his explanation of the "linear," for example, Wölfflin analogizes the movement of the eye with that of the hand in a perceiving subject's apprehension of an artwork:

> The tracing out of a figure with an evenly clear line has still an element of physical grasping. The operation which the eye performs resembles the operation of the hand which feels along the body, and

the modelling which repeats reality in the gradation of light also appeals to the sense of touch. (56)

Wölfflin's paradigmatic example of the linear is the drawing of Albrecht Dürer. The linear refers to a specifically oculomotor experience, the muscular movement behind the shifting of the eye in the apprehension of an object or a space, known as saccades (see, e.g., Walker and Doyle 2003). Because of the heterogeneity of the surface of the retina, this movement of the eyes with precise gaze control is essential for normal vision (see, e.g., Burr and Morrone 2004). Prefiguring such knowledge of the neurophysiology of vision in his definition of the "linear," Wölfflin's explanation of the "painterly" on the other hand abstracts the visual process back into the function of the eye:

> A painterly representation, on the other hand, excludes this analogy. It has its roots only in the eye and appeals only to the eye, and just as the child ceases to take hold of things in order to "grasp" them, so mankind has ceased to test the picture for its tactile values. (2003, 53–54)

So, whereas Wölfflin's example of the "linear" relies upon the specificity of the tactile organs in Dürer's draughtsmanship, the "painterly" deploys Rembrandt as a paradigmatic example to show a more generalized sensory intermingling, one which endorses the nonoptical or more-than-visual functions of the eye. "In Dürer everywhere the endeavour to achieve tactile values, a mode of drawing which . . . follows the form with its modelling lines" is eminently and immediately tactile, says Wölfflin, to be contrasted with Rembrandt's "tendency to withdraw the picture from the tactile zone and, in drawing, to drop everything which is based on immediate experiences of the organs of touch" (55). Riegl's identification of a loose historical shift from sculpture and bas-relief to more planar and painterly artistic forms, occurring particularly between Greek and late Roman art, similarly maps a shift from the direct appeal to the organs of touch to an expanded and multisensory notion of what the eye (and brain) can do. Shifting from the medium of painting to architectural experience, however, Wölfflin's "linear" is equated similarly with clear and distinct lines and edges that demarcate enduring forms, and yet that also involve inducements and movements of the eye:

> Linear vision ... means that the sense and beauty of things is sought first in the outline—interior forms have their outline too—that *the eye is led along the boundaries and induced to feel along the edges.*

whereas the painterly implies indistinct patches, making the forms look like something changeable, shifting, becoming:

> while seeing in masses takes place where attention is withdrawn from the edges, where the outline has become more or less indifferent to the eye as the path of vision, and the primary element of the impression is things seen as patches. (51, emphasis added)

Implicit within this formulation is an associated retinal movement and response to what is seen, the oculomotor responses and saccades present within embodied vision in identifying objects, perceiving outlines, and responding to forms or patches of color. Our eye is drawn around the objects in a painting, the lines and color patches, and enfolds other qualities into its perception.

While our eyes might be drawn or respond to certain features with particular movements, this in itself tells us nothing of aesthetic value, nor does it explain affective response. As viewing subjects are not purely optical beings, Wölfflin must look beyond the anatomical functions of the eyes, beyond the retinal surface: "Physical forms possess a character only because we ourselves possess a body," he declares (1993, 151). The physiological underpinnings of eye movement-sensations is only one aspect of spatial experience through the body that pertains to architecture. Like with Volkelt, for Wölfflin there is also a shift in architectural discourse from the usual inquiry into tectonic structures into the modes of spatial experience through the body, a more widely haptic rather than retinal proposition: "If architecture is the correlate of a lived experience, then this experience presupposes the movement of the body; the *kinaesthetic activity* of a body in motion justifies architecture as a practice that structures spatial relations," summarizes Ionescu (2016, 3). The beauty of an architectural space for Wölfflin is not merely a response to forms that are perceived on the retina, therefore, but the feeling of an unconstrained relation between the perceived forms and the "vital feeling" of our body. Hence Wölfflin returns to an idea previously raised by Volkelt, that of vital feeling *(Lebensgefühl).* This refers to a constancy of energy felt during the unconstrained movement of the body. It mediates between subjectively perceived forms in space, feelings of well-being, and the physiological structure of the body. The body of course is a necessary intermediary within any architectural encounter that potentially grants feelings of "well-being" *(Wohlbefinden)* or otherwise. Consequently, the body is not a passive receptacle of sensation but an original structure that we all inherently possess, and through which physical space is symbolically interpreted and our affective responses structured. This original structure is *"our bodily organization" (unsere leibliche Organisation),* says Wölfflin, *"the form through which we apprehend everything physical"* (1993, 157–58, emphasis in original). Wölfflin's premise, echoing Scott's words at the beginning of this section, is that the original architectural prototype is our own body. The physical properties of tectonic forms, of verticality and mass in a building, for example, are not just potentially pleasing to the eye but quantities that may be qualitatively felt as pleasurable. At one point Wölfflin uses richly introspective language in order to describe how

> Powerful columns produce in us energetic stimulations, our respiration harmonises with the expansive or narrow nature of space. In the former case we are stimulated as if we ourselves were the supporting columns; in the latter case we breathe as deeply and feeling as if our chest were as wide as the hall. . . . The architectural impression . . . is essentially based on a directly bodily feeling. (154–55)

The inherent motility of the body and the oculomotor nature of visual processes are inevitably involved when the viewing subject has experiences such as muscular contraction, responses to gravity, and interactions with materials. A more recent example of this idea is offered by contemporary architect and designer Bernard Cache. Given that the Baroque architectural style in particular makes use of gravitas and weightiness, Cache describes the mixture of bodily affect and perception that results: "This affect is as corporeal as it is spiritual: the heaviness of limbs as well as the preoccupations of the mind" (1995, 44). The reading of such characteristics within a built environment is dependent at least in part upon oculomotor processes such as the saccadic movements of the eye, like the prehensile hand, motile and exploratory. As Macarthur identifies, despite their oppositional stance regarding the place of *Einfühlung,* both Benjamin and Wölfflin "thought that our perception of architecture began in the body" (Macarthur 2008, 477). The constant movement of the immobile body is not simply a form of projection on behalf of the architect, or the beholder of a building, however:

> When architects design buildings, they do not imitate the static proportions of the human body, but also the mechanics of posture by which we hold ourselves erect, or recline under gravity. The body expresses movement even when motionless because that very lack of motion is achieved through the same interplay of skeleton and musculature that makes motion and extension into space possible. These poses and postures become a kind of language by which we can read the potential movements and emotive states of others. (481)

This is an apposite moment to concentrate on the relationship between stillness and movement of the eye, head, and body, mapping some significant neurophysiological connections between sensory subsystems regarding an ambulatory body when experiencing certain three-dimensional spaces.

Labyrinths, Vestibules, and Oculomotor Processes

In the course of many dissections of dogs and monkeys, the first-century Roman physician Galen correctly traced two sets of nerves into the ear, one from the muscles of the face and the other to the inner ear, and termed this inner region, after the Cretan myth, the "labyrinth" (Finger 2001). Vesalius's student Fallopius, in *Observationes Anatomicae* of 1561, revisited Galen's initial cursory treatment of the ear, separating out the three ossicles of the middle ear, or "tympanum," and divided the inner ear into two parts, the "vestibule" and its conjoining semicircular canals being the "labyrinth," while the neighboring snail shell-like structure he separated out as the "cochlea." The labyrinthine nature of the assemblage of the inner ear is a suitable allegory, as the snail-like cochlea, conjoined with three semicircular canals, and the "vestibule" that contains the bones that transmit sound to the auditory nerves, together involve dense spatial convolutions within the interior of the skull. For example, Duverney's illustration of the cochlea of 1683 clearly demonstrates a fascination with the spatial organization of the organ, but he misunderstood the function. Nevertheless, Duverney correctly understood the branching off from the vestibular nerve, one nerve going to the cochlear coil, another to the semicircular canal, and another to the vestibule (Swanson 2014, 161). For almost a century little changed in terms of representation and comprehension of the gross anatomy of the labyrinth in textbooks and treatises, so that Albrecht Von Haller's *Primae Lineae Physiologiae* (1747, First lines of physiology) similarly described the characteristic spatial convolutions, but also assumed the semicircular canals were involved in auditory processes (Wade 2003, 190). By the time Max Brodel's drawing of the inner ear was made in 1934, the true nonauditory vestibular functions were known (see Figure 9). Between these times, the discovery that liquid filled the inner ear was made and refined by Italian scientists in the eighteenth century, including Valsalva of Bolgna, Domenico Cotugno of Naples, who described the vestibule, semicircular canals, and cochlea of the osseus labyrinth of the internal ear, and Antonio Scarpa of Modena who, in *Disquisitiones anatomicae de auditu et olfactu* (1789), detailed his discovery of the membranous labyrinth with its "spiral passage," the fluid-filled cochlear duct (Hawkins 2004, 69). It was Flourens's experiments in the 1840s creating lesions in the semicircular canals of pigeons and then rabbits, and the later nonsurgical experiments by William James on deaf-mute humans (1881; 1882, see, e.g., Milar 2012) that conclusively demonstrated a nonauditory function of the semicircular canals. Flourens was the first to assign them the role of reflexive orientation, but made no connection between his experimental findings and the sense of movement (Wade 2004, 196). That was to be formulated in the Mach–Breuer theory of hydrodynamic semicircular canal function in the 1870s. Breuer's work on the sensory receptors of the inner ear culminated in the 1873 paper to the Imperial Society of Physicians in Vienna, "On the Function of the Semicircular Canals of the Aural Labyrinth," where he identified the role of the canals in sensing angular movements of the head, as a result of the movement of endolymph within the three planar orientations of the canals (Wiest and Baloh 2002). From precursors in amphibians and fish, Breuer inferred a similar mechanism for detecting linear acceleration in the utricle and saccule organs of the otolith, within the vestibule, that worked in synchrony with the semicircular canals, providing more comprehensive orientation information due to the sensing of movement. Furthermore, Breuer explicitly

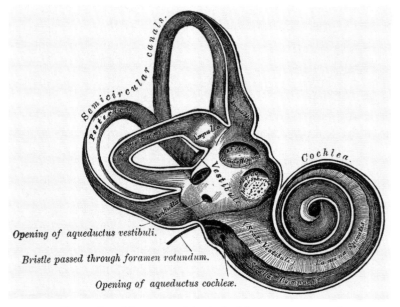

FIGURE 9. *Illustration of the osseous labyrinth of the inner ear, printed originally in the 1916 edition of Henry Gray's* Gray's Anatomy: Descriptive and Applied *(Philadelphia: Lea & Febiger, 1913), 1136.*

reconnected movement of the head to involuntary eye movements. Any angular rotations of the head, such as when spinning around, normally involve a rotation of the eyes in the opposite direction in order to maintain a visual fix on a distant object. Occasionally this reflex breaks down, leading to involuntary ocular oscillations known as nystagmus, or "dancing eyes." Therefore, as Wiest and Baloh explain, "Breuer was the first to explain the compensatory nature of the eye movement and its direct relationship to stimulation of the semicircular canals" (1649).

Contemporaneously, Ernst Mach was working on the anatomy and function of the inner ear in Vienna, with an unusual experimental setup to verify head rotation. In his *Fundamentals of Movement Perception,* he described what instigated his inquiry: "I noticed the tilting of houses and trees while I was traveling around a curve in a railroad. This is easily explained if one directly senses the resultant inertial acceleration" (in Wiest and Baloh 2002, 1649). Initially Mach postulated the illusion was explained by the "sensibility of the body as a whole" (in Henn 1984,

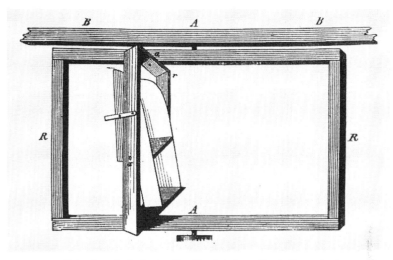

FIGURE 10. *Spinning chair devised by Mach to investigate the experience of motion. From Ernst Mach ([1875] 2001, 24). Illustration public domain, courtesy of SpringerNature.*

145) but later, like Breuer, reasoned that the sensors for such linear acceleration were located in the head. It is in his *Beiträge zur Analyse der Empfindungen* ([1896] 1914, translated as *The Analysis of Sensations*) that he talks of these types of sensations as "optical vertigo" (1914, 138). In a chapter titled "The Space Sensations of the Eye," Mach briefly discusses eye movements in appreciating aesthetic form, comparing rather simple line-drawn abstract figures to more complex Gothic architectural forms. This idea of the rapid movement of the eye responding to shapes harks back to Wölfflin's "kinaesthetic response of the eye," and builds upon Wundt's explanation of eye "movement-sensations." But Mach takes this much further. To prove his theory that the inner ear was responsible for sensing linear and angular acceleration, Mach famously constructed a chair device designed to rotate experimental subjects about multiple axes, with the head of the subject covered so that no visual input, and therefore no reflex ocular oscillation through nystagmus, be involved (see Figure 10). Visual aftereffects of bodily rotation remained, so he concluded that the semicircular canals sensed angular acceleration, not angular velocity.

Now that the vestibular function of the semicircular canals within the labyrinth had been discovered, alongside the auditory function of the cochlea, a brief summary of recent neurophysiological research on head orientation and eye movement shows the impressive legacy of this Mach–Breuer theory. Furthermore, at the risk of taking the neuroscientific perspective at face value, the phenomenon of the Vestibular-Ocular Reflex (VOR) will pick up from Wölfflin's ideas and filter into the phenomenologically inflected account of motile bodies in the final section below. Following on from the Mach–Breuer theory, the role of otoliths, the movement-sensing utricle and saccule that occupy the vestibule, is central. The otoliths sense the head's linear acceleration and are able to sense position relative to gravity. As the French neurophysiologist Alain Berthoz, who tested gravity effects on the vestibular system on astronauts, explains: "The otolithic organs (sacculus and utriculus) are inertial detectors of the linear acceleration of the head in the plane of their maculae. They also detect the angular displacement of the head with respect to gravity: when the head is tilted the otoliths are stimulated by the component of gravity in the plane of each macula" (1991, 91). Meanwhile, the semicircular canals detect rotational movement in each direction, as there are three canals oriented according to the usual spatial axes. Otoliths and canals very rapidly relay information to the spine and brain through the mechanism of reflexes, and one of the more prominent reflexes is known as the vestibular-ocular reflex (VOR) (see, e.g., Lackner and DiZio 2005, 119ff).

The extraocular compensatory mechanism that Breuer and Mach had identified above, to stabilize the retinal image of a static object during rotations or movements of the head, is the precursor of the VOR. In the language of vision science, the VOR removes "retinal slip" (see, e.g., Gebhart and Schmidt 2013). Through this reflex, any adjustments and movements of the body or head detected through the labyrinth will cause very fast and fine adjustments by the extraocular muscles that control eye direction. However, a fast-moving object, or indeed a fast-moving or fast-rotating subject, might fail to keep up with this gaze tracking, in which case "retinal slip" occurs. Retinal slip can be reduced through a rehearsed or practiced optokinetic adaptation, whereby the vestibulo-ocular reflexes are altered in order to produce "smooth pursuit"—that is, more correctly matching the movement of a tracked object through the movement of the eyes. Thus, with rapidly moving or rotating subjects, such as ballet dancers when they go "on point," the technique is to visually locate ("spot") a fixed object, such as a clock, continually maintaining eye contact with it while spinning (see, e.g., Osterhammel, Terkildsen, and Zilstorff 1968). Whether spinning, walking, or being stationary, smooth pursuit involves rapid compensatory use of extraocular muscles in relation to the orientation and movement of the subject's body or head. Berthoz underlines the significance of the labyrinth for this process, but rather problematically adopts a geometrical metaphor for the body's spatial position and orientation: "The semicircular canals are involved in stabilizing reactions. They are important either in reducing retinal slip, or in maintaining upright posture. They constitute an Euclidean reference frame for the measurement of the angular accelerations of the head," he explains (1991, 85). Notwithstanding his assumption of a Euclidian framework, Berthoz considers VOR as evidence of the functional parallels and reflex circuits between the semicircular canals and the oculomotor system, together providing a proprioceptive representation of the body's position in space. In fact, he persists in claiming that the conjoined sensory subsystems of posture and bodily movement entail a spatial reference frame: "The reference frame of the three semi-circular canals therefore has a fundamental role as a kind of template for the geometry of the sensory systems involved in the representation of movement" (89).

Rather than a geometrical approach based on a Euclidean spatial schema with representations of movement, one way of bringing together Breuer, Mach, and the neurophysiology of Berthoz is to return to Head and Holmes's (1911) original concept of "body schema"—that is, a non-Euclidean postural model of the body that deals with sensory impressions rather than representations, so modifies "the impressions produced by incoming sensory impulses in such a way that the final sensation of position, or of locality, rises into consciousness charged with a relation to something that has happened before" (Head, in Gallagher 1986, 542). A decade later, Schilder will equate Head and Holmes's postural body schema with the conscious image or representation based on sensory impressions of "our own body which we form in the mind" ([1935] 2013, 11), and equate the body schema to what he terms "body image." And a decade after that, Merleau-Ponty famously revisited Head and Holmes's body schema as a phenomenological means to understand the primacy of motility in the spatiality of one's body, and the incorporation of other objects into the body's habitual set of motoric skills

(Merleau-Ponty 1992; see also Paterson 2018a). I have already mentioned the body schema in the introduction, and a more sustained examination in relation to Merleau-Ponty's fundamental concept of "motor intentionality" will take place in chapter 6. For now, however, I stop short of the phenomenological method sometimes appealed to in recent aesthetics and architectural theory, including Pallasmaa, Holl, and Perez-Gomez. Instead, despite his problematic use of the term "representation," Berthoz also favors the idea of "simulation": "Perception and simulation of action are embedded in each other" (1991, 24). This concept simultaneously refers back to one of the earliest motor theories in science, that of R. H. Lotze in 1952 with his idea of *nachahmungsbewegungen* (a tendency to copy or imitate perceived motions, like in empathy), but also forward to more enactivist or sensorimotor theories of perception such as that of Alva Noë and Kevin O'Regan. In the following final section, therefore, I bring together some of the neurophysiologically informed aesthetic, architectural, and art history ideas encountered so far through Mach, Breuer, and Wölfflin in order to consider oculomotor processes, and the kind of empathic simulation of perception in encountering a complex built environment by way of a conceptual "walkthrough," or case study.

Motility of Eye and Feet: From Sculpture to Mosaic
In this section I turn briefly to a consideration of the spatial art of sculpture, whose textural and dimensional qualities are enhanced through an ambulatory encounter, and then more fully an archaeological site. The movement of the beholder is clearly fundamental to the appreciation of sculptural form. Herbert Read's controversial postwar contribution *The Art of Sculpture* (1956) argued directly for an aesthetics rooted in sculpture's physicality, what he called its "ponderability." For Read, modern sculpture was to be understood as "a three dimensional mass occupying space and only to be apprehended by senses that are alive to its volume and ponderability, as well as to its visual appearance" (1956, ix). Echoing the earlier discussion of Riegl, Berenson, and Wölfflin on painting, and the limitations of retinal vision, for Read the physicality of sculpture's material properties worked as a counterpoint to what he termed the "visual prejudice" that had plagued criticism of sculpture until then. If the encounter with the sculptural work remained predominantly visual, something like a haptic sensibility was acquired in the

apprehension of the sculptural object's physicality, its ponderability, weight, and volume. Indeed, in Read's appreciation of sculptural form from pre-Colombian times right up to modern sculpture, the defining orientation was one of an imagined hapticity, since all sculpture was considered "an art of palpation—an art that gives satisfaction in the touching and handling of objects" (49). And, against that sense of the interplay of volumes and shadows on the surface that Le Corbusier had articulated, Read's argument directly concerned the specificity of the medium, and seemingly alludes to Johan Gottfried Herder's influential *Das Plastik* ([1778] 2002, translated as *Sculpture*):

> The specifically plastic sensibility is, I believe, more complex than the specifically visual sensibility. It involves three factors: a sensation of the tactile quality of surfaces; a sensation of volume as denoted by plane surfaces; and a synthetic realization of the mass and ponderability of the object. (71)

This mode of criticism now seems quaint. Yet what he calls the "synthetic realization" of those qualities, derived (at least in part) from complex ocular processes, speaks directly to the oculomotor nature of a perceiving subject. In Read's rejection of a purely optical interpretation of sculpture, there is nevertheless a production of more-than-visual values from complex ocular processes. Something like this is also true of religious relics, as Susan Stewart discusses. In eighteenth-century French salons, for example, illustrations show mobile and seemingly unruly spectators being closely intertwined with sculptural works, "turning them almost into extensions of their own bodies" (1999, 29), before the more familiar disciplining of moving bodies in museum spaces and the "do not touch" rule is instigated (see, e.g., Leahy 2012). Such hitherto unregulated proximity complicates this division because there is already a heightened physicality that emphasizes the sculpture's sensual presence, inviting movement around—and even the touching of—nude sculptures until the 1780s. Even within current museum settings, of course, the mobility of bodies remains crucial, as Stewart reminds us:

> A key difference between the temporal immediacy of visual perception and the spatial immediacy of tactile impressions is the latter's motility. To experience the roughness or smoothness of an object,

to examine its physical position or come to understand its relative temperature or moistness, we must move, turn, take time. Visual perception can immediately organize a field; tactile perception requires temporal comparison. (Stewart 1999, 32)

The fixed point of visual perception, that single, immobile point near the center of a work that is optimal for a visual encounter since Alberti, as we saw, entails a distinct and privileged location for the viewer to reclaim its intended spatial effects. But, as art historian Rebecca Molholt finds in the case of a Roman floor mosaic, its meaning "was inseparable from its experience as a tangible surface, one typically appreciated by an ambulatory viewer situated in and aware of a specific architectural setting" (2011, 287). Rather than a detached opticality, or an imputed tactility, real or virtual, of sculpture, these Roman forms and materials were designed to be situated, and therefore felt, underfoot, to be appreciated kinaesthetically.

Built between the second and fourth centuries BCE, two examples of floor mosaics from Roman settlements in North Africa, Belalis Maior and Thuburbo Maius, are now housed in the Bardo Museum in Tunis. The complex of Thuburbo Maius is a site fifty-three kilometers to the southwest of Tunis in the fertile Milania Valley. According to Jashemski, Thuburbo was made into a *municipium,* a tribal center under Roman control, under Hadrian, and became a colony under Commodus (1995, 561). It was first excavated in 1915 by the French director of the Antiquities Service in Tunisia between 1906 and 1920, Alfred Merlin. Shortly after the initial excavation, Merlin relates how his team had completed the clearing of a large bathing area, the *thermae aestivales,* or summer baths: "En 1915, nous avions achevé le déblaiement d'un vaste établissement balnéaire, les thermae aestivales" (Merlin 1917, 67). A dedicatory inscription found by Merlin showed that the bath complex was completed in 361 BCE (Bates 1919, 188). Then, less than a year after the initial discovery, a further bathing area of around 1,600 square meters was discovered that, based on an inscription from a nearby forum, was most likely the *thermae hiemales,* or winter baths. Mosaics were found throughout the settlement, including in the gardens of private houses (Jashemski 1995). But larger, more ornate mosaics were unearthed around the bathing areas that were richly decorated with statues of the gods but also, according to Merlin, light-colored mosaics that brightened the floor of most rooms ("des mosaïques aux couleurs claires égayaient le sol de la plupart des salles" [1917, 71]). In attempting to read and translate inscriptions on lintels with weathered or damaged stone, Merlin describes a struggle with one reference to the construction of a swimming pool with the word "cochlea." At first he wonders whether this was the name of a machine used to elevate the water level, mentioned in fact by Vitruvius, but then conjectures that the pool was nicknamed "the snail," the animal after which the inner ear organ is named. Even were this so, why would this be?

> Why this strange name? Is it because of the shape of the pool? Is it because of its mosaic decoration or a carved ornamental pattern representing a snail? Nothing in the excavations has come to enlighten us on this point. The inscription does not seem to distinguish by the word cochlea a special kind of swimming pool, but to use this term as a current nickname attributed to a swimming pool for a cause that escapes us. (74, my translation)

Such impressive bathing areas would be part of a larger public gymnasium with a palaestra, an athletic school that taught boxing and wrestling. The mosaics here were not simply floor decoration, but accessible spaces with an iconography designed to involve the visitor in a wider bathing context, a heightened physicality and athleticism around the dedicated wrestling rooms, and the warm and well-decorated central bath (tepidarium), around which both hot and cold bathing rooms were located. For, while young athletes were trained in the rooms, scholars and the general public would also meet for discussion in the larger spaces and also bathe. Book 5, chapter 11 of Vitruvius's *De Architectura* of 27 BCE is devoted to the palaestra, a building he described as common among the Greeks as well as the Romans, and a structure that involves a walkway of two stadia in length. Writing of the palaestra, Vitruvius advises future builders to "construct roomy recesses . . . with seats in them, where philosophers, rhetoricians, and others who delight in learning may sit and converse" arranged around the central "young men's hall" (Vitruvius 1914, 160) where the physical training was sited. It is no accident therefore that Plato uses the palaestra as a setting for two Socratic dialogues, and metaphors that explore a heightened physicality in the struggle for truth recur throughout the Platonic dialogues. In the *Phaedrus* the conquest of the baser part of the soul by the virtuous part is compared with a wrestling victory in the Olympic games (*Phaedrus* 256b), but a discussion of wrestling in the *Protagoras* differentiates wisdom from strength (*Protagoras* 350e). The palaestra also features in Plato's *Lysis* (206b), where Socrates is drawn into a discussion of *Eros* (sexual love) versus *Philia* (friendship) during a festival where young and older men gathered together. In Plutarch's *Alcibiades* the palaestra of Sibyrtius is the place where a follower of the titular philosopher was killed.

It is recorded that most visitors to the public bathing areas exercised first, and only bathed "after a good sweat had been worked up" (Fagan 1999, 10). The arrangement of the rooms reflects this bodily regimen, the mosaic floor design reflecting this heightened physicality by inviting the eye, and also the foot, into ambulatory narratives involving familiar and heroic mythic episodes or scenes. Covering large swathes of floor in the bathing complexes, scenes of aggressive boxing, or the labyrinth at Knossos with figures of an athletic Theseus and a vanquished Minotaur at the very center, both drew the eye and invoked a muscular physicalism in the viewer (see Figures 11 and 12). "Walking across these spaces helps to construct the bather as an heroic athlete," claims Molholt, since "the beholder is moved to become actively engaged in the narrative unfolding underfoot" (2011, 287–88). Quite how engaged with the strikingly physical narrative a visitor can be is revealed in A. J. Liebling's 1956 essay for the *New Yorker,* "An Ancient Thuburban Custom." "On the floor of the Bardo Museum in Tunis, there is a mosaic picture of a knockdown in a prizefight that took place about 200 A.D." (in Thompson 1956, 79), the essay starts, the better to contextualize the actual boxing match in Tunis he would later report on. Liebling, a long-running boxing journalist, neatly conveys the dynamism of the scene, the bodily forces at work, but also how astonishingly recognizable and realistic it seems to modern eyes. The Thurburban fighter is standing over his bearded opponent, having given the knockout blow, blood from the left temple spurting from the point of impact represented by a sequence of red-colored mosaic pieces. Liebling writes of the bodily position of the standing fighter delivering the knockout punch to his bearded opponent: "His left foot, slid flat along the ground, is in advance of his right, but not too far, and he is up on the ball of his right foot as he throws

FIGURE 11. *Labyrinth Mosaic from Thuburbo Maius, Tunisia. License: Creative Commons.*

FIGURE 12. *Theseus Mosaic, discovered in the floor of a Roman villa at the Loigerfelder near Salzburg in 1815, fourth century AD, Kunsthistorisches Museum Vienna, Austria. Photograph: Carole Raddato from Frankfurt, Germany, CC BY-SA 2.0.*

the punch, with all his body in it" (Liebling, in Thompson 1956, 80). As for his opponent: "The older fighter is squatting on his hunkers, neither knee quite touching the ground. The punch has not dazed him; he has his elbows pulled in tight to his body and his fists in front of him, ready to hit as soon as he can bounce up" (Liebling, in Thompson 1956, 80).

Initially, Roman floor mosaics would seem to confirm Le Corbusier's explanation of the static retinal properties of shapes and volumes, as their flatness is a requisite property for their architectural setting. The aspect of physicality for the bather, the use of these rooms as exercise spaces prior to bathing, is further enriched through the variation of floor temperatures and water temperatures around the various rooms and baths. Like gymnasiums, unsurprisingly, bathing spaces were also designed for a heightened physicality. Given their place within the larger physical context, with sculptures, masonry decorations, and architectural features such as porticos, floor mosaics are designed to be encountered differently from wall paintings. Figures depicted in the floor's center were to be viewed from the side, having some rudimentary three dimensionality from this orientation. It is the ambulatory viewer, the oculomotor subject of heightened physicality, who responds in a transitive way, moves, and then becomes absorbed into the familiar mythic narratives depicted. Given the areal expanse of the mosaic, the labyrinthine designs draw the eye separately from the entirety of the ambulatory body, and, on first encounter at least, this may be halting, arresting, might temporarily inhibit or disturb movement. The difficulty lies in following any one pathway with the eyes and so, at particular junctures, a rift between opticality and hapticality may open up as the eyes become drawn away from the feet: "The spectator is optically lost, even though [s]he is standing on an open floor," imagines Molholt (2011, 290). This does not negate the idea that motility is already present or implied in the composition of the mosaic, a motility that entwines the eye of the viewer with the inherently spatial narrative that even a schematic labyrinth offers. Especially after walking past, or on, mosaic depictions of athleticism and heightened physicality, therefore, "the beholder has stepped into the role of the hero since his own movement, both visual and physical, implicates him in the myth underfoot," thinks Molholt (295). This is indeed a different haptic sensibility, a texture felt through the foot and halted, drawn, jolted by the eye. It is not simply due to different forms of texture underfoot, those piecemeal fragmentary obtrusions that constitute mosaic. Instead, there is a different attentional filter in terms of pedestrian perception.

Earlier, Gibson's definition of the "haptic system" involved a multimodal sensory apparatus, the means by which animals and men are "*literally* in touch with the environment" he said (1968, 98, emphasis in original). It is this literality of contact that motivates anthropologist Tim Ingold, in his article on perceiving the world through the feet, to consider a "more literally *grounded* approach to perception" (emphasis in original), revaluating the place of touch in the balance of the senses: "For it is surely through our feet, in contact with the ground (albeit mediated by footwear), that we are most fundamentally and continually 'in touch' with our surroundings" (2004, 330). A neglected, yet extremely commonplace form of haptic perception, Ingold's phenomenologically-influenced approach intends this pedestrian touch as a corrective to the concentration on manual touch in social science (although see Paterson 2009 for discussion of "haptic geographies"). In so doing, however, he assumes a normative body, one without missing limbs, sticks, canes, the wheels of a wheelchair or, in increasingly automotive cultures, of a car, in *feeling* grounded. Ingold's pedestrian haptics therefore echo Merleau-Ponty's *Phénoménologie de la Perception* and Gibson's haptic system in the supposed generality for all human subjects, then, as there are no culturally specific bodies, unlike fellow anthropologist Marcel Mauss's closer analysis of the specific *habitudes* of bodies and their styles of walking, swimming, and so on (Mauss [1934] 1992). After presenting a typology of such movements that the ethnographer could find articulated in their cultural specificity, Mauss admits that such variations presuppose "an enormous biological and physiological apparatus" that links the social and biological body through a "psychological cogwheel" (474). But this cogwheel is one of many, the implication being that distinct styles and expressions of movements are irreducible to solely psychological explanation, and so are "biologico-sociological phenomena" (474). Despite these concerns, Ingold's revisitation of a distinctly pedestrian haptics is assessed in this section in terms of its capacity for advancing the project of an embodied, mobile perception, an ambulatory mode of bodily encounter, that may remain amenable to the forms of oculomotor subjectivity described above. Ingold questions whether the usual perceptual categories of figure and ground, central to models of vision and even manual touch, still pertain to pedestrian touch (and consequently, I would add, to the more distributed conception of haptics or somatosensation from the second section, "Haptic Perception"). Are there particular qualities and sensations of pedestrian touch that are distinct from manual touch? He continues to question:

> What difference does it make that pedestrian touch carries the weight of the body rather than the weight of the object? And how does the feel of a surface differ, depending on whether the organ of touch is brought down at successive spots, as in walking, or allowed to wrap around or slide over it, as can be done with the fingers and palm of the hand? (2004, 330)

The ambulatory subject walking through the bathing complex is certainly undergoing a heightened physicality, but one that has already been accomplished within the setting, since the ball of thread that was left by Ariadne for Theseus finds its representation, in each instance of the mosaic, already at its center. As a labyrinth, therefore, it is not only a design but also an optical puzzle that draws the beholder in, incorporates the spectator into both a visual and an ambulatory narrative. This narrative is additional to the already heightened physicality of walking and looking in museum contexts that Stewart had noted. In *Museum Bodies,* Leahy takes up Kirshenblatt-Gimblett's observation that "mobility is a defining feature of the museum," yet notices how little attention has been paid to "the felt quality of its navigated space" (in Leahy 2012, 74). In addition, the fundamental way that "walking choreographs visuality within the museum," as Leahy neatly puts it, is mostly ignored in favor of that previously noted curatorial imperative for bodies to remain immobile in front of paintings or artifacts (2012, 75). Following Mauss, there are clearly social and spatial "bodily techniques" involved in ambulatory viewing, rhythms that respond to the spatial setting and the presence of other visitors that accrue over time. For ambulatory visitors in the mid-nineteenth century, thinks Leahy, the combination of "walking, introspection and self-awareness" cultivated a heightened self-presence, an intensification of the senses akin to the flâneur (79). With such biologico-sociological considerations in mind, I return to another form of heightened physicality, that aggregation of tactile experiences and muscular movements Read described, above, to consider the transposition or projection of the ambulatory beholder of the mosaic.

With no optimal angle or approach for the flattened yet textured design, it must be negotiated through movement, scanned with the eyes, felt with the feet, and in some form synchronized through previous proprioceptive experience in order to obtain some form of narrative immersion, to make the encounter resonant:

> Across the terrain of these labyrinth floors, it is the perceiving subject who finds, in the flat floor underfoot, a stage for a narrative that elevates his own movement and an invitation to synchronize his actions with the work of art. Visual perception is just one layer of this process of bodily perception and projection: depicted space complements and expands actual space via the introduction of history, imagination, and mythology. Motion is . . . the key to the labyrinth mosaic. (Molholt 2011, 296)

To move, in other words, through an arrangement of spaces with textures underfoot, the eye being drawn in a linear rather than painterly fashion, the physicality of the beholder being heightened through an agonistic culture of athleticism enshrined by mythic encounters. The eyes moving relative to the orientation of the head, with bundles of reflexes between the vestibular system of the inner ear and the extraocular muscles; and the rapid saccades tracing out the marks of a labyrinth on the floor, stimulating separate, yet connected, ocular and proprioceptive pathways. Such is a description of the oculomotor subject in action.

This chapter started with an acknowledgement of some limitations of phenomenology for architectural theory, and in proceeding it has engaged with a range of academic fields and time periods in order to piece together a narrative of moving eyes and motile bodies. Throughout this exploration of the oculomotor subject, the types of tactility and haptics that rely on the hands, on manual interactions with objects, has been deliberately offset here in favor of reconsidering the place of quotidian multimodal experiences of eye movement, vestibular orientation and reflexes, and perception through the feet. Historically, theories of vision almost inevitably trump considerations of haptics, especially vestibular and pedestrian sensations through the feet. This tendency is what Ingold pejoratively terms the elevation of "head over heels" in Western culture (2004, 315ff). It is tempting to directly counterpose the seeming certainties of retinal vision, identified with organs located in the head, against the distributed and sometimes indistinct sensations arising through the body, routed through the labyrinth, the extraocular muscles, and stretch receptors in the muscles of the body. I have argued that, on the one hand, it is something like these (seemingly) barely coherent oculomotor responses and muscular movements that take their part in what Mauss has termed "biologico-sociological phenomena." On the other hand, bodily movements do kick-start the process of cognition, claims Ingold:

> Locomotion, not cognition, must be the starting point for the study of perceptual activity.... Or more strictly, cognition should not be *set off* from locomotion, along the lines of a division between head and heels, since walking is itself a form of circumambulatory knowing. (331)

Between this more generalized and potentially normative way of attending to pedestrian haptics of Ingold, or the more socially and culturally specific types and styles of movement and bodily techniques that Mauss and Leahy attended to, it is clear that the neurophysiologically aware concept of the oculomotor body is not straightforwardly biologically "given." Unlike Ingold, I do not cleave forms of cognition associated with the "head," nor the retinal vision associated with the eyes, from a loosely determined aggregation of bodily sensations through the feet. Instead, a historically informed reading of the neuroscience, of a more intertwined relationship between the vestibular system and oculomotor behaviors, offers a set of resources for critically interrogating this assumed opposition between head and heels, or between visual and haptic systems. This chapter therefore marks a modest starting point for further investigation along these lines, to advocate an attention to forms of perceiving and apprehending an environment premised on more than mere bodily movement or locomotion. Oculomotor processes such as rapid eye saccades, and reflexes that are wired into vestibular subsystems, occupy a space in the architecture of the body between physiology and culture, forming both an inside and an outside of conscious control and awareness in terms of perceiving the environment.

The intention of my narrative has been to help update and extend proto-phenomenological approaches to experiencing architecture and the built environment by drawing on the art history, physiology, and psychology of Mach, Breuer, Wölfflin, and others for the productive nexus of neurophysiology and the theorization of aesthetic experience. The case study of the bathing complex is rather rarified, and more everyday architectural spaces could easily have been chosen instead. Yet, as both optical phenomenon and texturally rich, underfoot, experience, the floor mosaics and surrounding complex work as a sensorily immersive environment, akin to far later entertainment spectacles such as the painted panoramas of the eighteenth century, or the theatrical dioramas of the early nineteenth century. Such examples would equally serve to demonstrate the heightened physicality of oculomotor subjects in environments that invite ambulatory activity. Through the case study, I hope to have shown a point of departure, one pedestrian step further toward considering how art history, architectural theory, museum studies, and other disciplines might benefit from prior historical articulations of current thinking on the body and bodily sensation.

CHAPTER 4

"The Neuro-motor Unconscious"
Étienne-Jules Marey, Eadweard Muybridge, and Motion Capture

J. M. W. Turner's 1844 painting *Rain, Steam and Speed—The Great Western Railway* is justly celebrated for its proto-impressionistic portrayal of a symbol of modernity, a steam train, emerging out of a blurred atmosphere of lashing rain mixed with the roiling steam pouring from the locomotive's engine. A review of the National Gallery's exhibition from the *Times* of London singles out the painting. It shows the locomotive "in very sudden perspective, and the dark atmosphere, the bright sparkling fire of the engine, and the dusky smoke, form a striking combination" (*Times* 1844, 7). The wall note beside the painting includes the barest of descriptions: "A steam engine advances across a bridge in the rain. In front of the train, a hare runs for cover" (in Thomas 2016, 18). The hare, as Turner probably knew, was the fastest animal native to Britain and consequently a recognizable index of speed. Thematically, the painting is held to be elegiac, at once affirming speed and progress while mourning the displacement and loss of the natural world, a celebration of "old England and new England combined," argues the sociologist Francis Donald Klingender in his *Art and the Industrial Revolution* (in Carter 1997, 4). Yet the immediate effect on the viewer of 1844, a whole fifty-one years before the Lumière Brothers' early film "Arrival of a Train at La Ciotat," must have been similarly startling. "There comes a train down upon you," wrote the poet William Makepeace Thackeray, Turner's contemporary, after seeing the painting at its first hanging in the National Gallery in London. "The viewer had best make haste . . . lest it dash out of the picture and be away to Charing Cross through the wall opposite" (in Thomas 2016, 15). The landscape, rather than being a sun-dappled and static pastoral scene, consists almost entirely of an admixture of the dynamic forces of nature and the progress of man-made industrial technology. The one area of relatively sharp definition in the otherwise rough brushwork, that as it were, seems "in focus" to the viewer, is the smokestack and glowing firebox at the leading edge of the black Firefly locomotive as it slices through the haze. An optical effect now familiar to us through photography, the simulated depth of field seems to offer the viewer an impression of movement captured in time. Despite singling out the painting within the exhibition as being "striking," the *Times* reviewer's summation of Turner's contributions to art remained inconclusive. "Whether Turner's pictures are dazzling unrealities, or whether they are realities seized upon at a moment's glance, we leave his detractors and admirers to settle between them," writes the anonymous reviewer (*Times* 1844, 7).

Needless to say, it was not just the steam locomotive that seemed to burst suddenly out of the landscape of modernity and industrial progress at this time. For the first photographic images had already been circulating, including daguerreotypes from 1837. Some satirical instructions in the Parisian magazine *Charivari* for August 30, 1839, for taking a daguerreotype portrait revealed the well-known limitations of the method for capturing living subjects, since any movement would degrade the clarity of the image. To make a portrait of your wife, for example: "You fit her into a fixed iron collar to give the required immobility, thus holding the world still for the time being. You point the camera lens at her face; but, alas, you make a mistake of a fraction of an inch [in focusing], and when you take out the portrait it doesn't represent your wife—it's her parrot, her watering pot—or worse" (in Newhall 1976, 27). For now, the fixed image required fixed subjects. But while some subjects might assent to being fixed in space temporarily, the rest of the world was moving ever more rapidly. From 1840 onward a series of camera patents and chemical processing innovations followed in quick succession, including the United States' first camera patent for Alexander Wolcott's modified daguerreotype box camera in May 1840, William Henry Talbot's patented Calotype process in 1841, and Frederick Scott Archer's Collodion process in 1851, each successively reducing exposure time to eventually capture fixed images permanently on paper. Rather ironically then, and irrespective of the rapid progress in photography at the time, the means to depict "realities seized upon a moment's glance," as the *Times* reviewer so elegantly put it in 1844, would be through painting for a while longer.

As the barely visible hare in Turner's painting sped ahead of the furious metal locomotive, the mysteries of animal locomotion would be revisited thirty years later, as advances in those photographic and chronophotographic techniques and processes renewed interest in the topic, this time with a range of instruments and techniques for the truly scientific observation of elusory movement. Giovanni Borelli's *De Motu Animalium* (On the movement of animals) of 1680 was a respected treatise on biomechanics whose title referenced the text of the same name by Aristotle of around 335 BCE.[1] As influential as Borelli's text was, hypotheses on the contraction of the muscles and the effects of fatigue on movement were empirically unverifiable and remained in the realm of speculative physiology. How the biomechanical study of locomotion becomes revitalized through the rapid development of photographic processes and new optical apparatuses is the starting point of this chapter. For, in a key period between 1872 and 1886, an unusual concatenation of quite distinct forces and motivations occurred involving an American industrial capitalist and later politician, an eccentric Englishman whose claim to fame was photographing the landscapes of the American West, and a French physiologist who devised ever-more ingenious apparatus for the study of animal, then human, locomotion. The messy drama of their mixed collaborations, rivalry, lawsuits, and subsequent falling out not only plotted a path to empirical verifiability of movement patterns through photographic means, but also irrevocably altered the painting of animal movement in the academies, and the perception of that movement in the eyes of the public. Toward the close of that extraordinary century, the experimentalism regarding movement and the capturing of images in time became a profoundly new area of philosophical study, courtesy of Henri Bergson.

This is a chapter of three broad sections, each corresponding roughly to the way that two near-contemporaneous figures seek to capture, and then conceptualize, animal and human movement in the wild. Although their job titles, geographical provenance, and route to prominence differ markedly, this was a truly interdisciplinary pursuit. Eadweard Muybridge was a journeyman artist whose work stimulated the scientific community. Étienne-Jules Marey was an eminent Parisian physiologist with an original and applied scientific methodology who ended up irrevocably impacting the visual arts. Remarkably, Muybridge and Marey were born the same spring of 1830, and died a week apart from each other, in May 1904. But whereas Muybridge was a dilettante, a man of serial self-reinvention who discovered at one point the scientific

necessity of photographing animal locomotion, Marey's mission throughout his career remained focused on the centrality of movement: "Not long
ago the origins of movement, that is to say, muscular action and nervous
function, were the most mysterious aspects of biology," Marey stated
in 1868 in *De mouvement dans les functions de la vie,* his collection of
lectures at the Collège de France. Now, however, "as a result of the contributions of the German school" of physiology they would be "best
understood" (in Rabinbach 1992, 87). The first section, "Realities Seized
upon a Moment's Glance," examines the nine-year association between
Muybridge and his sponsor Leland Stanford at his stables in Palo Alto
from 1872 onward, fueling experiments with photographic images of
motion sequences of horses that in 1878 helped definitively reveal the
single instant when all four hooves left the ground. This is a familiar-
enough narrative, yet the value of this section is in setting up the correspondence between Muybridge and Marey that occurred from 1878
onward and their eventual meeting in 1881. Although there was no
actual collaboration between the talented but eccentric artist and the
methodical and resourceful scientist, there was certainly mutual inspiration and the exchange of ideas through meetings and written correspondence. Marey's *La Machine animale* of 1873 (translated as *Animal
Mechanism* in 1874) was unmistakably an influence, with its tracing
and diagramming of the movement of parts of bodies through an extension via photography into what Marey called the "graphic method,"
first formulated in *La Méthode graphique* in 1878 (as yet untranslated),
but whose principles were applied and continuously modified throughout his scientific career. As a result, photography suddenly played a
crucial role in indexing and verifying the movement of animals and
men beyond the realm of normal human perception.[2] The second section, "The Animal Machine," therefore takes up the longer arc of Marey's
experimental physiology both before and after Muybridge, including
an exploration of the formation of the graphic method. This involved
collaborating with others, including the engineer Charles Fremont, to
adapt the design of existing measuring instruments into more portable
versions that could be used in the field. Decades before the physiologist Charles Sherrington left his laboratory to study the movements of
its workers and the impact of fatigue on the production line of a wartime munitions factory (see chapter 5), Marey began to extend his
physiological research on animal and human locomotion outside of the

laboratory. The extension of the spaces of observation into what we will term "extramural physiology" ultimately leads him to build a specialized Station Physiologique, a physiology station, with tracks, instruments, and stables in the grounds of Paris's Parc des Princes in 1881, with the help of the photographer Georges Demenÿ. In the third section, "Analysis and Synthesis of the Biomechanics of Movement," the cumulative narrative of Marey's contributions to the physiology of locomotion reveals one of the central problematics of this chapter—namely, the illusion of the concrete and objective form of time consciousness through chronophotography as endlessly divisible and therefore subjectable to analysis, yet which is experienced subjectively as synthesis through the combination of the perceptual apparatus (the eye, along with technologies such as the zoetrope, the zoopraxiscope, and the magic lantern) and the function of memory. In other words, here I examine how movement captured through chronophotography is reconstructed as moving image or rendered as dynamic model, subject to further analysis through diagramming, what Marey will term "geometric" or "partial" chronophotography.

Realities Seized upon a Moment's Glance: Muybridge and the Photographic Capture of Motion

The potential of photography to arrest movement and then animate it as a sequence of individual frames was pursued by Eadweard Muybridge and others as so-called instantaneous photography. But the limitations of early photographic chemical processes and apparatuses encouraged a robust experimentalism in the capture and portrayal of movement that predates this. The Collodion process from 1851 necessitated a long exposure time with an emulsion or "wet plate," but even such fixed limitations led to a slew of creative approaches to producing an illusion of movement through photographic series in the 1850s. Experiments by André-Adolphe-Eugène Disdéri and Antoine Claudet are mentioned by Philip Brookman, the former producing "images of people and animals interacting with each other while moving," and the latter creating "a sequence of sequential self-portraits showing the illusion of action through time" (Brookman 2010, 78). In 1853 Claudet patented a stereoscopic viewer to better depict such illusory movement, "for pleasing and novel optical illusions . . . which are made movable so as to impart to the picture the appearance of moving figures" (Claudet, in Brookman 2010, 78). In the 1860s Louis Ducos du Hauron, who over his lifetime took few photographs himself, nevertheless conceived of two big photographic innovations. In 1868 he patented a new technique using three different-colored glass filters as negatives that, when put together, produced a composite positive color image, explained further in his monograph *Les couleurs en photographie, solution du problème* of 1869. Earlier, in 1864, du Hauron was granted a French patent for the earliest known motion picture apparatus that, in a single camera, had two sets of sixteen lenses arranged side by side. "By means of my device, I make myself able particularly to reproduce the passing of a procession, a military review . . . the grimaces of the human face. . . . This will be in some fashion a living representation," he declared (in Winston 2005, 237).[3] There is no record of the device actually being built. Meanwhile, Darwin's *The Expression of the Emotions in Man and Animals* (1872) included large plates with photographic sequences of facial expressions charting how different emotions registered on the faces of human subjects in a series of grids. That very year began the fruitful collaboration between Muybridge and the tycoon, governor of California, and later senator Leland Stanford, whose stables had quickly become a "laboratory for motion," in Rebecca Solnit's words (2004, 190). On the back of Muybridge's reputation with photographs of landscapes such as Yosemite National Park's fast-flowing waterfalls and vast granite mountains, Stanford cabled Muybridge to invite him to photograph his horses. Muybridge moved into a cottage on Stanford's eight thousand–acre Palo Alto estate, close to the private racetrack, to work on the technology to capture the horses' motion. The following year, 1873, brought together a number of related endeavors. Darwin's photographer, Oscar Gustave Rejlander, published an essay "On Photographing Horses," proposing the use of "a battery of cameras and 'quick acting lenses' ready charged and loaded" (in Brookman 2010, 78) to produce a sequence of images depicting motion. That year Muybridge successfully photographed a horse in motion called Occident owned by Stanford, although the photographs were not widely seen by the public. In fact, alternative media such as lithographs, engravings, and even paintings were reproduced from those photographs. It so happened that 1873 also brought public attention to the study of locomotion, the science of movement, and the study of flight through the publication of Étienne-Jules Marey's *La Machine animale:*

Locomotion terrestre et aérienne, soon translated into English as *Animal Mechanism* (Marey 1874). Muybridge, who had already made something of a name for himself by publishing photographs of landscapes from his extensive travels through the American West, returned his attention to photographing horses in 1876. The actual episode of the multicamera motion capture of Stanford's galloping horse eventually published in 1878 is now familiar, especially regarding its contribution to the birth of cinema (see, e.g., Braun and Whitcombe 1999; Solnit 2004; Brookman 2010; Latsis 2015). However, my purpose in this section is to establish how Muybridge's contribution to the analysis of movement through photographic apparatuses was taken up as a more extensive physiological project, first by himself and then with others including Marey, as they enjoyed a short correspondence and subsequent meeting in Paris in the wake of a profile of Muybridge's experiment in *Scientific American* in 1878.

Rebecca Solnit claims that Muybridge's work with Stanford's patronage and interest produced three breakthroughs in three consecutive years: the year 1877 produced a single image of a horse in motion with a single mounted "automatic electro-photograph"; 1878 involved mounting successive images for motion rather than a single image through an array of cameras; then 1879 brought the further innovation of mounting the successive images within an apparatus of Muybridge's own invention, a zoopraxiscope, to reanimate the photographic sequence and therefore produce the first "moving picture" (Solnit 2004, 185). For the first breakthrough in 1877, Muybridge had already been working for several years to reduce exposure time and therefore "freeze" the horse's motion more effectively. This involved refinements with the chemical process and experiments with a trigger and rubber springs to snap wood planks together in front of the lens to constitute a remotely operated mechanical shutter. Prior to this innovation, photographic exposure was managed by manually removing and then replacing the lens cap. As Muybridge explained to a reporter for the *San Francisco Bulletin* on August 3, 1877: "These experiments resulted in the construction of an apparatus and the preparation of chemicals so as to permit the photographing in outline of a rapidly moving body" (in Brookman 2010, 80). It was from a photographic session in July 1877 that Muybridge first successfully achieved his aim to "stop motion" of a galloping horse with all hooves off the ground through a single camera using the Collodion process, triggered by a thread strung across the track. Muybridge found the resulting single image disappointingly indistinct, and it was copied and retouched within a watercolor and gouache painting for reproduction in newspapers. This was not unusual, as photographs were often enhanced for the sake of clarity, as Brookman reminds us (2010, 79), and translated into another medium to facilitate their wider dissemination. It was one thing capturing the single image of motion, however clear and distinct as a result of these innovations. To understand horse locomotion more fully, a series of such images must be captured.

Further experiments with arrays of cameras housed in a shed by the side of the track helped produce sequences of motion with six and then twelve frames that were subsequently printed and sold. The June 1878 version of the photographic equipment produced the most decisive and celebrated sequences. For, that month, a series of photographic cards were produced that showed Stanford's horses Abe Eddington, Sallie Gardner, Mohomet, and Occident variously trotting, cantering, and galloping along the track. Rather than a single camera, this time there was an array of cameras in the manner Rejlander originally envisaged, but with a series of wires every twenty-one inches that ran across the track to the shed, each wire acting as a trigger for an electrical circuit. Along with Muybridge's innovations in chemical processing, an engineer on his team named Issacs made a dramatic improvement to the shutter mechanism, conceiving of a system whereby the movement of a springed magnetic shutter could be triggered by electricity. As the horse ran it tripped the wires, pulling the trigger that closed the electrical circuit for each camera in the array, and releasing rubber springs loaded at one hundred pounds of pressure that snapped the shutters closed at one-thousandth of a second. The press were invited to a demonstration of the system on site in June 1878, where small glass plates were developed in situ to dispel any doubts about what was captured and then shown to reporters. A few months later, the sequence "The Horse in Motion," featuring a very stylized painted series of Abe Eddington, was reproduced on the cover of *Scientific American* of October 19, 1878, with the tagline: "The science of the horse's motions." The accompanying article explained how Muybridge had perfected an "automatic electro-photographic apparatus" and succeeded in "recording the action of horses in motion" (1878, 241). The exposure time was two-thousandths of a second, far removed from the painfully sluggish daguerreotype, and this was only possible due to a coordinated burst of technological leaps and photochemical innovations. Of course, the media coverage focused on the charismatic ringleader: "Mr. Muybridge employed a series of cameras, operated by electricity, and so placed as to fix with absolute accuracy the several phases of the continuous action of the horse while making one stride," explains *Scientific American* (241). The illustrations definitively revealed that the horse's hooves were all suspended from the ground when galloping at a gait of 2:24 or faster.

Muybridge's third great breakthrough followed shortly after the famous stop-motion sequence and was something he initially termed a "zoogyroscope" before settling on "zoopraxiscope." In fact, it was a short step from creating sequences of stopped animal motion to their being placed within the existing technology of the zoetrope, or zootrope, also known as Plateau's Phénakistoscope, a spinning cylindrical device that had been around in some form since the 1830s, and first patented in 1867, into which was inserted a paper strip with successive images of movement. As it had regular slits cut out of the vertical side, the viewer was presented with a flickering illusion of continual movement as long as the device kept spinning. As another *Scientific American* article suggested: "Everybody is familiar with the zoetrope, an instrument which, when set revolving, portrays some moving figure—e.g., a horse in full gallop. Formerly the pictures—each of which represented a different stage of movement—were drawn by hand, but now by the introduction of photography the zoetropic representation of motion has been brought to a beautiful degree of perfection" (1896, 344). Although usually considered an entertainment device, Marey had previously seen the value of the zootrope in *La Machine animale* (1873) for the study of locomotion, as it "presents to the eye a series of successive images of persons or animals represented in various attitudes. When these attitudes are coordinated so as to bring before the eye all the phases of a movement, the illusion is complete; we seem to see living persons moving in different ways" (1874, 137). He continued:

> This instrument, usually constructed for the amusement of children, generally presents grotesque or fantastic figures moving in a ridiculous manner. But it has occurred to us that, by depicting on the apparatus figures constructed with care, and representing faithfully the successive attitudes of the body during walking, running, &c., we

might reproduce the appearance of the different kinds of progression employed by man. (137)

The progression of human and animal gait would be a convincing illusion at normal speed, but a great advantage, thought Marey, was the ability to slow down the device, "so that the eye can ascertain with the greatest facility these actions, the succession of which cannot be apprehended in ordinary walking" (137). This suggestion of a marriage between photographic sequence and zoetrope clearly fired the imagination of others, and attempts to bring this to fruition spread far and wide. The *Scientific American* article in 1878 had urged such a possibility for Muybridge for the same reason: "By such means it would be possible to see not only the successive positions of a trotting or running horse, but also the actual motions of the body and legs in passing through the different phases of the stride" (1878, 241). One year later, Émile Duhousset wrote from Paris that he had successfully adapted Muybridge's photographs and mounted them in an old phenakistoscope, as had W. B. Tegetmeier in London in June that year, and the painter Thomas Eakins in Philadelphia "had plotted . . . the successive positions of the photographs and constructed, most ingeniously, the trajectories" that were then adapted into bands for mounting in the zoetrope (Rogers, in Haas 1972, 27).

As if to further cement this wish, another high-profile scientific report of Muybridge's success occurred in the French journal *La Nature*, which prompted Marey to write a letter to the photographer via the publication, which included the following: "I am impressed with Mr. Muybridge's photographs published in the issue before last of *La Nature*. Could you put me in touch with the author? I would like his assistance in the solution of certain problems of physiology too difficult to resolve by other methods" (in Solnit 2004, 195–96). Just as the *Scientific American* article did, Marey similarly recommended the reanimation of stopped motion through the zoetrope, although not only as a scientific aid for the study of animal movement, as he had expressed earlier in *La Machine animale,* but also as a representational aid for artists: "Then, what beautiful zoetropes he could make. One could see all imaginable animals during their true movements; it would be animated zoology. So far as artists are concerned, it would create a revolution for them, since one could furnish them with true attitudes of movement; positions of the body during unstable balances in which a model would find it impossible to pose" (in Solnit 2004, 200). In the same letter, he implored Muybridge to point his camera lenses toward flying birds. Muybridge's great innovation therefore was not just to bring together the cylindrical zoetrope with his photographic sequences, but to incorporate another technology he had used on his lecture tours, the magic lantern. Versions of the magic lantern, an apparatus that used a light source to project through a painted-glass screen, had been around since the seventeenth century, and since 1849 had been used to project photographs.

Within a month of the 1878 sequence of the horse in motion, Muybridge had already used the magic lantern to project a short sequence of life-size images "in quick succession" to simulate movement in his lecture tours, as Mozley explains: "The near-synthetic effect of one slide quickly succeeding the other must have made him determine to devise a way to project them in quick enough succession to correctly reconstitute the motion his instantaneous photographs had stopped" (1972, 71). The zoo-praxiscope was a step further, a combination of his own photographic sequences of images, the zoetrope, and the magic lantern. It used a glass disc upon which a number of hand-painted silhouettes, and later colored images, were drawn, with a counterrotating slotted disc geared to spin at equal speed, which had the same intermittent viewing effect as the zoetrope. This assembly was connected to a projection lantern that could cast almost life-size images on the walls. In 1870, a projection device with a complex wheel of slide holders called a Phasmatrope was demonstrated by Henry Hyle in Philadelphia, rotating a projected sequence of images that offered the illusion of movement (Winston 2005). Had du Hauron's motion picture apparatus worked, the image sequences would have been perfect for Hyle's projector a decade earlier, but here was a combination of motion sequence and projector in Muybridge's hands for the first time. Perhaps because of precursors like du Hauron, Muybridge was unable to patent his zoopraxiscope device either in America, England, or France, so could not officially claim to be its inventor. Nevertheless, he clearly saw the apparatus's potential for "synthetically demonstrating movement analytically photographed from life," as he put it (in Haas 1972, 24). This had also been the case for Muybridge's earlier photographic sequences, of course, as the potential for the truly scientific study of animal locomotion had prompted the *Scientific American* cover story and Marey's letter to Muybridge to express excitement about their value for physiology, as well as the possibility of mounting them in a zoetrope. Through a zoopraxiscope, projected as much larger moving images on the walls of theaters, Muybridge's motion sequences could now be viewed by a larger audience, with even greater entertainment value. Of course, this novel apparatus constituted a vital step in the birth of cinema. The reanimated loops of the zoetrope projected by the magic lantern on the walls of lecture theaters enthralled the public with what seemed startlingly lifelike, as well as life-size, to those audiences. When Muybridge showed his photos of horses in this way at the San Francisco Art Association in May 1880, a reporter from the *San Francisco Daily Call* marveled at what the zoopraxiscope was capable of, describing the projected images as "apparently the living, moving horse," and that "nothing was wanting but the clatter of the hoofs upon the turf and an occasional breath of steam from the nostrils to make the spectator believe that he had before him genuine flesh-and-blood steeds" (in Ball 2013, 21). Just like his first static images of 1873, and unbeknown to his audience, however, Muybridge had to ask an artist to paint facsimiles of the photographs onto the glass disks and, according to Edward Ball, to give the horses "elongated legs and torsos, counteracting the optical distortion effects" (2013, 326) of his new zoopraxiscope. Just like his earlier celebrated photographic sequences that captured movement, then, to appear clear and lifelike for his audiences the images had to be artistically enhanced. This approach will differ markedly from that of Marey, as might be expected.

Anita Mozley was the curator of a 1972 retrospective of Muybridge's photographs at Stanford University, "Eadweard Muybridge: The Stanford Years, 1872–1882." Her introduction to the catalog emphasizes the centrality of movement as the basis for Leland Stanford's collaboration with Muybridge, although the myth of a bet between the tycoon and the photographer about the position of hooves in the air during the horse's gallop has been hard to dispel. Leland Stanford's estate became a "laboratory for motion" for a reason, as a "theory of animal locomotion was what Stanford was after" (Mozley 1972, 9). Muybridge's collaboration with Stanford came to an acrimonious end, one reason being that Stanford continued his pursuit of locomotion through the study of horses by commissioning another collaborator, the medical doctor J. D. B. Stillman, to write a volume entitled *The Horse in Motion* (1882). Stillman asked Muybridge to contribute an introduction to the book, which was submitted but never used, and made line draw-ings from Muybridge's photographs that were. The resulting court case of *Muybridge v. Stanford* brought up testimony by Stillman that under-lined Stanford's original motivation and his genuine interest in equine locomotion. Speaking of his wealthy patron, Stillman testified: "It was a hobby with him to explain the phenomena of motion. He always said that nobody understood anything about it, and that he was determined to unravel the mystery of motion. He could not explain it himself, but he was satisfied that all explanations were unsound" (in Solnit 2004, 78). Even if Muybridge's original photographs were transformed into line drawings in Stillman's volume, the following section will show how Marey also transformed the supposed veracity of original photographs of locomotion through the "graphic method"—that is, converting them from a shadow-like naïve realism into a series of abstract graphic rep-resentations, the better to depict, and subsequently analyze, the inter-mediary steps of human and animal movement. Meanwhile, Muybridge continued his pursuit of photographic motion capture in alternative, sometimes rather curious, contexts. His photographic sequences of animal locomotion had interested the University of Philadelphia for the potential insight it offered in the fields of sports, medicine, and physiology. It was there that Muybridge created more than twenty thousand photos for his eleven-volume publication *Animal Locomotion: An Electro-photographic Investigation of Connective Phases of Animal Movements* (1887). This impressively extensive study had a bewilder-ing variety of photographic subjects, including women lifting bedsheets, raising cigarettes to their lips, and a rather whimsical thirty-six-picture series *Crossing Brook on Stepping-Stones with Fishing-Pole and Can* (plate 175). Taken from three different angles, the sequence follows a woman as she raises her leg, hops from one stone to another, then hops off, for some reason holding a fishing pole in one hand and a can in the other. The mechanics of her movements are the main subject, with her ankles, knees, shoulders, and elbows rotating along their individual joints, her weight shifts evidenced by the sequence of muscle contractions. Although idiosyncratic in composition, these sequences were used by artists such as Edgar Degas and Marcel Duchamp to study motion. Duchamp's 1912 painting *Nude Descending a Staircase, No. 2* is visually indebted to Muybridge's photo series *Woman Walking Downstairs*, re-printed as plate 121 in a subset of his original publication that became

The Human Figure in Motion (1955, 243–44). A very clear and regular grid with a black background and white horizontal and vertical lines features in a number of Muybridge's photographic sequences, including *Woman Walking Downstairs.* An almost identical grid would become prominent in photographs of human movement in Frank and Lillian Gilbreths' scientific management literature early the next century (see chapter 5). This photographic grid had previously appeared in the anthropometric photographic studies of Malaysian males by John Lamprey in 1869. Such grid systems were apparently "overt expressions of the positivist concentration of the 'mathematicization of empirica,' and the related notion that images, like graphs, could work without text and become a controlled lexical space," according to Edwards (in Braun and Whitcombe 1999, 224n19).

The following section dives further into the mechanisms and instruments of scientific motion capture. However, since this chapter started with the artistic representation of the movement of a train, and Muybridge himself was not averse to the creative alteration of photographic images, a brief coda. Another contemporary of Muybridge and Marey was Jean-Louis Ernest Meissonier, famous at the time for his masterful portrayal of the gait of a walking horse in *French Campaign of 1814* (1864). There is a curious effect in the wake of the discovery of photographic verisimilitude that breaks with, or perhaps simply breaks, prior artistic convention. For, once Meissonier had been exposed to Muybridge's motion sequences, apparently the galloping of horses became impossible for him to paint. Take the first four of the sixteen photos in the famous sequence *Sallie Gardner at a Gallop,* for example, from the June 19, 1878, session, which shows all four hooves off the ground. These photos look nothing like any previous artistic depiction. According to a story in the *San Francisco Examiner* in 1881, Stanford met with Meissonier in Paris and requested: "Sketch me a horse trotting." The artist obliged. Then Stanford requested the painter to sketch the same horse in stride a foot later. Meissonier tried several times, taking a step back each time before rubbing out the lines. Finally, he admitted: "I can't do it" (in Musser 2005, 19).

The Animal Machine: Marey and the Physiology of Motion

Three years after his letter to Muybridge was published in *La Nature,* and just as Muybridge's relationship with Stanford was fully unraveling,

Étienne-Jules Marey would host Muybridge at a salon in his house in Paris in September 1881. Also present was Hermann von Helmholtz and Nadar, pseudonym of photographer, essayist, and balloonist Gaspard-Félix Tournachon. Muybridge did bring the bird pictures that Marey requested, but they were not the scientifically rigorous motion sequences he expected. Marey expressed his disappointment that "these were not representative of successive attitudes, as they were for the man or the horse" (Marey 1882, in Rabinbach 1992, 103). Nevertheless, he continued, "if it were possible to obtain the images in series," as Muybridge had achieved so remarkably with his horses, it would validate the photographic method for the study of the mechanisms of flight. Nevertheless, their meeting was transformative in terms of photographic technique, as "Marey's involvement with photography dates from this period and clearly draws from Muybridge," observes Solnit (2004, 210). Specifically, because of Muybridge's photographic sequence from 1878, Marey was now aware of the potential of the introduction of the dry plate into photography and how this might extend the possibilities of the graphic method, as Marta Braun details. Muybridge had shown him that photography "has a greater role to play in science; it allows one to approach extremely complex problems and give a concrete solution with a singular facility," thought Marey (in Braun 1994, 47). The same year that they met in Paris, Marey adapted astronomer Pierre Jules César Janssen's existing photographic revolver, which captured images of an eclipse on a rotating glass plate, into the celebrated *fusil photographique*, or "photographic gun," which debuted a year later, in 1882 (see Figure 13). If Muybridge had fixed the locomotion of animals and then humans in space through photographic sequences, Marey will further fix this movement in successive instances of time through what he terms *chrono-photographie*, and then further abstracts the movement through "partial" or "geometric" chronophotography, illuminating only the movement of particular body parts within the chrono-photographic frame by means of silver strips and spots against a black background to aid analysis. The tireless inventiveness of Marey, and his single-minded devotion to studying the biomechanics of movement, would have far-reaching impacts on technology and culture, involving a continual dance between the mechanographic, the physiologic, and the aesthetic.

FIGURE 13. *Marey's* fusil photographique *of 1882, published in his paper in* La Nature *that year (Marey 1882). Illustration in public domain, from Wikimedia Commons.*

Physiology as a scientific discipline was still comparatively young in the first half of the nineteenth century. François Magendie had founded the first laboratory for experimental physiology in France at the Collège de France, and in 1821 established the first journal in that field, *Journal de Physiologie Expérimentale.* Claude Bernard became a laboratory assistant at the Collège under Magendie before his own appointment as the first professor of physiology at the Sorbonne. Both Magendie and Bernard were eminent scientists, but physiology was traditionally seen as subservient to anatomy, and rarely treated seriously in medical teaching. At the midpoint of the century, when Marey moved further into his physiological studies after his medical internship with Dr. Martin Magron, the status of physiology was more solid, buoyed by Helmholtz's work on thermodynamics and the physiology of the senses and Carl Ludwig's graphic method. From the 1850s the growth in observation, registering, recording, and graphical inscription of life phenomena meant a "laboratory revolution" was underway in Germany, as Sven Dierig (2003) shows. Physiology gained in reputation and indeed became a *Leitwissenschaften* (lead science), or as Emil du Bois-

Reymond put it in an opening ceremony of his new institute in Berlin in 1877, "the queen of the natural sciences" (in Dierig 2003, 118). According to Dierig, du Bois-Reymond had himself gained inspiration from Carl Ludwig's Physiological Institute in Leipzig, which opened in 1869, "the most prominent physiological institute in the late nineteenth century and the first to follow a factory-like working organization" (122). Just like a factory, large machines delivered mechanical power to individual workshops, only the vivisection table was substituted for the vice, with a comparable fixing of the animal-object in place.[4] Meanwhile, in France the first clinical laboratories opened in 1873, and until then physiology could only be taught by individuals with the means for private laboratories and equipment, according to Braun (1994). By the time Marey was conducting his own research, the physiological laboratory was a site inextricably associated with scalpels and vivisection, where destructive acts of static dissection were a prerequisite for any empirical observation and analysis. When experimental physiologists did have access to laboratories, their dismal craft usually involved dissection. Wilhelm Wundt stated unequivocally in his *Lehrbuch der Physiologie* (1852) that vivisection "was the first step on the ladder toward knowledge of an organism's interior world," says Dierig (2003, 128). The irony that, to understand how movement occurs in living organisms, they had to be stilled, euthanized, and then dissected was not lost on Marey. He thought such laboratories were "sad, dingy and unhealthy rooms in which researchers damn themselves to live in the mere hope that they may someday discover the characteristics of cell structure and the functions of the bodily organs" (in Mayer 2010, 103). Wishing to distance himself from them, since his appointment as chair of "Natural History of Organized Bodies" (following Flourens's retirement) also at the Collège de France in 1869, Marey had made forays outside of his rooms to conduct experiments, including a large lecture hall within the university to study the movement of birds and insects, visiting various riding stables around Paris, and jostling with the public at the Jardin de Luxembourg.[5] The annoyance of traveling so continually "was but slight compared with the grave disadvantages of carrying long distances delicate instruments whose slightest derangement made the journey useless," wrote Marey. This was not a long-term solution, and so "finding a spacious ground where I might unite a workshop, laboratory, and an experimental field" (1896, 397) became a priority. A purpose-built "physiology station" was necessary to further pursue these extensive extramural physiological studies. His vision would come to full fruition in 1881, but in the meantime, there are false starts and, all the while, a set of adaptations of existing instruments and apparatus to study movement, and after Ludwig's and Helmholtz's contributions, the continual refinement of his own distinct "graphic method."

Marey's impressive and sustained research in the natural sciences seemed to occupy a space between an inventiveness for the mechanization of instruments of capture and a perpetually unsated curiosity regarding the mechanisms of animal, and at various points human, movement. Marey dedicated himself to studying the movements of the body in the same way as Borelli's *De Motu Animalium,* in terms of biomechanics. His fascination with machines of all types, his manual dexterity, and his various collaborations with engineers led him to the study of human mechanics, Braun argues, "the study of the levers, pulleys, pumps, pipes, and valves within the living body, and of the forces that drive them" (1994, 11). In *La Machine animale* of 1873, translated as *Animal Mechanism,* Marey explicitly makes this parallel, claiming that "the laws of mechanics are applicable to animated motors" (1874, 59). The history of the idea of the "animal machine" far precedes this, of course, but had renewed power during the laboratory revolution in physiology. For example, Du Bois-Reymond used the term *thierische Maschine* (animal machine) in numerous speeches (Dierig 2003, 129n83). Marey started with internal movements, studying the circulation of the blood and the movements involved in respiration, and collaborated extensively in this phase with the veterinary physiologist Jean Baptiste Auguste Cheauveau to invent sophisticated instruments to graphically inscribe these formerly imperceptible processes, such as the wearable sphygmograph (1859), the polygraph (1863), and used the sphygmograph to produce cardiographs of heart movements in a series of joint articles between 1861 and 1864. In 1868, with the publication of *Du mouvement dans les fonctions de la vie,* Marey remained committed to Helmholtz's myograph, originally from 1849, adapting it to study the movements involved in muscular contraction and to measure the speed of nerve impulses. Key to these new means of measurement and inscription was his version of the "graphic method," as initially outlined in a publication of that name in 1878. By the 1870s Marey turned his attention to external motion, abandoning his interest in internal movements, "cardiac cycles, ventilation, the twitch of a frog's leg at the application of an electric impulse," in favor of "the running of the horse, the flight of birds, walking, jumping and human gesture" and later still the movement of fish and flies, explains Dagognet (1992, 65). With his usual industry and inventiveness, and with the collaboration of engineers, he attached mobile mechanical devices to the bodies of horses, men, and birds as documented in *La Machine animale*. As discussed below, Marey "constantly sought to imagine instruments that could go beyond the thresholds of perception and transcribe phenomena themselves (autography, the language of nature)," Dagognet summarizes (39).

Even if Eadweard Muybridge's multicamera photographs of animal movement in England were inspiring but not scientific enough, Marey nevertheless saw great potential in the photographic method for the biomechanical analysis of movements. Acknowledging Muybridge's contribution, yet also its limitation to an external perspective of motion, Marey argues:

> The analysis of the movements of man and animals may be made from different points of view. It is, indeed, not sufficient to determine the external characters of the movement; the important matter is to ascertain the mechanism by which that movement is effected and to distinguish the part played by the different portions of the locomotor apparatus, muscles, articular surfaces, and osseous radii. (1896, 399)

The brief article "Chronophotography" in the *Scientific American* of 1896, for example, reminds us that Marey "was among the first to elaborate the chronophotographic method and to extend it to fields of interest in medicine" (1896, 344). Noting the publication of Pritchard's English translation in that year of Marey's book *Le Mouvement,* and how its subject matter should be of general interest, the article continues: "Chronophotography has also afforded fresh information of a most important and interesting kind as to the nature of physiological movement," even the smaller-scale movements involved in vital activities within the bodily interior such as cardiac movements or the tracking of blood in capillary vessels. Along with the chronophotographic method of tracing vital physiological processes, Marey also adapted and miniaturized von Vierordt's sphygmograph as a more portable metrical instrument for the circulatory system wearable on the arm as part of his doctoral thesis in 1863, and applied his investigation of animal and human lo-comotion to methods of gymnastic and military training, reported, for example, in *La Machine Animale* (1873) and *Le Mouvement* (1894). Along with these monographs, Marey wrote for popular scientific publica-tions in his native France such as *La Nature,* but summaries of his re-search were also reported within English-language publications, with *Scientific American* regularly featuring brief reports and even trans-lating some of the short articles from *La Nature.* The true breadth of Marey's research becomes apparent.

As a natural scientist with a keen and rigorous approach to experi-mental observation, Marey therefore produced photographic appara-tuses such as the *fusil photographique* of 1882 not just to "capture" lo-comotion in animals and humans but to produce a spectral streak or trace of motion more representative of the underlying biomechanical processes taking place. Further, by departing from the naïve verisi-militude of motion that appears in Muybridge's artistically retouched photography, Marey could escape what François Dagognet calls the "prison of the retina," combining various instruments and mechanisms of capture with the graphic method to discover a new territory, "the neuromotor—made of rhythms, muffled pulses and fluxes traversing the corporeal machine (producing nervous discharges, reactions: in short, the automatic writing of nature itself)" (1992, 132). Marey's in-struments and chronophotographic technique were being developed at a time when internal physiological phenomena of movement, rhythm, and pulse were increasingly subject to outward measure and optical rep-resentation, where the tendency was "to regard the animal body itself as a matrix of surfaces," according to Robert Brain (2008, 402). Lorraine Daston and Peter Galison find the mid-nineteenth-century fascination with graphical recording devices an instance of the ideal of "mechani-cal objectivity" (1992, 84), or "the mechanical construction of pictorial objectivity" (115), a form of objectivity resting on the exclusion of any potentially weary human observer. The psychophysicists of the first half of the nineteenth century who measured sensations, such as Ernst Heinrich Weber's work on touch and pain (see chapter 2), aimed for such objectivity but nonetheless relied on the presence of an assistant to apply the instruments to portions of the subject's body as well as to manually record the results. Marey's great contribution was "the discovery of how to make recordings without recourse to the human hand or eye," writes Dagognet, encapsulating Marey's worldview: "Nature had to testify to itself, to translate itself through the inflection of curves and subtle trajectories that were truly representative. Hidden, minute and fleeting, life's movements had to be captured (life is movement and nothing else)" (30). Whether through direct inscription based on bodily contact, or as a more distal chronophotography that initially lacked precision and therefore required aesthetic enhancements, Marey saw the graphic method as a way to produce objectivity through machine-made images of nature ("the automatic writing of nature itself"), where the instruments would themselves be the observers, and where automation and authenticity were valued as an almost moral imperative for scientists to approach "truth to nature."

CURVES, PULSES, AND WAVES: MAREY AND THE GRAPHIC METHOD
As if to directly address the problems inherent in the measurement of sensation by psychophysicists such as Fechner and Weber, and their assumed objectivity, early in *Le Méthode graphique* (1878) Marey outlined his distrust of sensory experience within the scientific process. The senses could be instructive, but were also deceptive. Observation equipment that already successfully minimized the role of subjective sensation in other scientific disciplines such as astronomy showed the path:

> The spherical nature of the earth, its daily rotation, the distances of the stars and their immense volumes, all our knowledge of astronomy, so to speak, contradicts the appraisal of our senses. The same can be said for a whole host of notions in physics and mechanics, such as the heaviness of air, the discontinuity of sounds and light, and so forth. The sensations of coldness and heat that our sense of touch provides no longer have the absolute meaning they were once given. (Marey 1878, in Dagognet 1992, 17)

The basic components for a continuously recording graphical measuring device were assembled by Thomas Young as early as 1807 in what he called a chronograph, which involved a vibrating metallic rod that inscribed regular wave patterns onto paper wrapped around a rotating drum (Rabinbach 1992, 96). Mechanisms and instruments that involved more direct bodily contact, and which inscribed curves onto drums and paper as a codified system, were experimented with in medicine, physiology, and the arts particularly between 1847 and 1930. Most famously this included Carl Ludwig's kymograph (1847) for correlating arterial pressure and thoracic air pressure, Hermann von Helmholtz's myograph (1849) for measuring muscle contractions, and Karl von Vierordt's sphygmograph (1854) for writing the pulse. Marey, write Daston and Galison, "dreamed of a wordless science that spoke instead in high-speed photographs and mechanically generated curves; in images that were, as he put it, in the 'language of the phenomena themselves'" (1992, 81). The tracing of curves, lines, waves, and pulses therefore featured prominently in this wordless language of nature. Ludwig is regarded as the originator of the graphic method, and his kymograph for measuring arterial pressure was "the first self-recording instrument to be introduced in physiology," according to Soraya de Chadarevian (1993, 267n2). The kymograph was aptly named from the Greek *kūma,* "swell" or "wave," as it was designed to chart the relationship between respiration and blood circulation. Variations in pressure were transmitted via a vertical stem to a cylinder rotating at a consistent speed.[6] "The curve could be retained and considered as continuous evidence," says Dagognet, and because of this direct sampling "the transfer onto paper allowed objective, verifiable examination as well as exact measurement" (1992, 29). The directly inscribed flowing curves that resulted from the technical solution Ludwig derived "was soon perceived as nothing less than 'a symbol' of the new disciplinary growth of physiology," writes de Chadarevian (1993, 268–70). At no less than the opening of the Physiological Institute in Berlin in 1877, Emil Du Bois-Reymond, who knew Ludwig's work, stated: "The conception of the quantities of so-called life-phenomena as a function of variables, and the, as it were, *bodily inscription [leibhuftige Aufzeichnung]* of their course of action in curves, has led to a completely new treatment of old problems" (in de Chadarevian 1993, 284, emphasis in original). The trend of modifying and then refining existing instruments and repurposing them for graphical output did not start with Marey, as even Ludwig had adapted an existing instrument of Spengler's, and Helmholtz had refined his own myograph to measure the velocity of nerve impulses, for example. Yet the range of graphical inscription mechanisms that Marey brought with him to his lecture series "Du movement dans les function de la vie" at the Collège de France in 1868 was impressive, including a cardiograph, a pneumograph, and a portable polygraph along with the kymograph, the myograph, and his version of the sphygmograph. In applying such *appareils inscripteurs,* or inscribing instruments, across a whole spectrum of species and activities, Marey further developed his "graphic method" as described in *La Méthode graphique:* "The graphical method translates all these changes in the activity of forces into an arresting form that one could call the language of the phenomena themselves, as it is superior to all other modes of expression" (1878, iii, in Daston and Galison 1992, 116).

The graphic method was broader than any single empirical methodology, a meta-methodological project influenced by the positivism of Comte, wherein photography and especially chronophotography could be co-opted into its overarching and universalizing project of graphical abstraction and diagramming. In the *Archives de physiologie* of 1889, Marey explained how photographs could be used to reveal a "natural expression which is not that of ordinary language: the latter is too slow and too imprecise to describe clearly, with their complexity and variability, the different acts of life" (in Canales 2009, 79). If language was an imprecise and limited means of conveying movement, then perhaps other already-existing notation systems might be up to the job, Marey considers in *La Méthode graphique*: "For a long time already . . . we have commanded a graphical expression of very fleeting, very fine, and extremely complex movements, which could never be expressed in a language. This admirable writing is being read in all countries, it is, in the true sense of the word, a universal language. I refer to the musical notation" (in de Chadarevian 1993, 277). Marey, a friend of both Helmholtz and Ludwig, believed that only through the graphic method could physiology become an exact science akin to physics, and expressible in mathematical form, and as Braun puts it "that by subjecting life processes to ever more scrupulous measurement, the numerical relations that underlie their manifestations could be determined with precision" (1994, 12). The resulting graphical inscriptions would be universal in two ways, as Daston and Galison summarize:

> Graphical representation could cut across the artificial boundaries
> of natural languages to reveal nature to all people, and graphical
> representation could cut across disciplinary boundaries to capture

phenomena as diverse as the pulse of a heart and the downturn of an economy. Pictures became more than merely helpful tools; they were the words of nature itself. (1992, 116)

The promise of direct measurement was central to the graphic method, and if instantaneous photographs and the interstitial capture of continuous movement through chronophotography showed animals and humans in previously unrecognizable postures, nevertheless "many scientists believed that these photographs might overcome the limitations of preexisting scientific imaging techniques," writes Jimena Canales (2009, 78) in her history of scientific research on nervous transmission.

With the help of the engineer Charles Fremont, Marey's chronophotographic techniques were not only applied to vital subdermal processes but also, as in Muybridge's later career, to photographically document the bodily movements involved in artisanal skills, the better to improve labor practices and to streamline automated assembly lines as a precursor to the Taylorism that would sweep Europe and America around 1913. "The motive power of the human body" was the title of one such report, itself taken from a paper Marey presented at the International Congress of Hygiene in 1884, where he described a means of measuring this motive power for a range of movements through precise mechanical and pneumatic means: "Planks, with India rubber coils underneath, recorded, by expelling the air they contained, the exact pressure of the foot" (*Scientific American* 1884, 290). Such a means of measuring force through displacement of the air combined Marey's aptitude for problem-solving with the potential for practical outcomes since, if the same force exerted by a walker were increased in frequency, then the distance covered would be far greater. Marey recommended the pace from forty to an optimal seventy-five steps per minute. This combination of scientific earnestness with an implicit striving for greater efficiency, along with the use of graphical forms and method in the observation of workers, would be revisited in later research on muscular fatigue with implications for industrial workplaces, explored in the following chapter. Clearly, chronophotography offered a new method of rendering unseen movements more visibly and therefore constituted another avenue to gauge efficiency:

Chronophotography, combined with the use of dynamometers, provides us with exact information about all of the acts we execute, which often reaches our consciousness in a very incomplete form. In this way it becomes the educator of our movements, allowing us to recognize the ideal perfection that we should attain and to recognize both our incorrect movements and the progress that we have made. Thanks to the progress of the graphical method, the mechanical act of locomotion can be translated into geometric curves, in which everything becomes measurable with a precision that mere observation could never attain. (Marey, in Mayer 2010, 88)

Despite Marey's fervent and persistent dedication to the graphic method, some of his contemporaries were not so enamored. This included criticism of Marey's assumptions at the very heart of his project, "the pretension that the graphic method should be the starting point of all exact and certain science" (in Canales 2009, 71). This was especially pertinent given that the measurement of reaction time of experimental subjects, and various attempts to gauge the speed of nervous transmission, were in contention, with apparatus and methodologies involved by leading scientific figures of the day such as Charles-Edouard Brown-Séquard and Hermann von Helmholtz attempting to objectively measure the speed of transmission of sensory and motor nerves. The scientific methodologies were hotly contested, but there was a growing acknowledgment of the influence of "psychic stimuli," the internal subjective state of the experimental subject, and whether they were fatigued, well exercised, or even suffered inflammations of the spinal cord from disease, say, impacted on the results. As Canales points out, the growing uncertainty among the scientific community regarding any assumed objectivity within the measurement of nervous transmission certainly made them suspicious of Marey's claim of direct measurement through his graphic method. More worryingly, the way that Marey had assumed graphical instruments produced consistent and universal legibility was now subject to question. What about the variability of the curves produced across these devices, and to what extent was this variability due to limitations of the instruments themselves? For example, Marey was aware of the potential for variability in rotational speed of the drum around which the paper formed a recording surface, or the variation in speed of muscle contraction between individuals. Marey's petition that the instruments generate "real curves" *(courbes reélles),* he rather fancifully suggested, might therefore have to involve an international commission just like the one established for weights and measures (de Chadarevian 1993, 288–89). The language of graphs was interpretable only to those who were aware of the conventions and limitations, and even if the graphic method was supposed to be a direct measurement of physiological movement and therefore transparent, it still involved reading or interpreting the inscriptions that resulted from essentially another codified system. Such objections and doubts within the scientific community led to a "crusade" against the graphic method by members of the Académie de Médecine, for example, pursued by various authors through the pages of its journal the *Bulletin de l'académie de médecine*. A contributor to the *Bulletin* in 1878, the same year as Marey's key publication, wrote, for example: "The registering apparatus does nothing to inscribe undulating lines that fall on our senses; but once it comes to interpreting the traces, the graphic method has no more certitude than direct observation," and continued damningly: "If everyone can produce these traces, not everyone is capable of interpreting them" (in Canales 2009, 71). Elsewhere in the *Bulletin* that year, Gavarret echoed those sentiments: "In the sciences of observation, all instruments, no matter how simple or complicated, are aids . . . that speak a special language. Before using them, one must strive to learn their language" (in Canales 2009, 71–72). The conventions, the ability to interpret this graphical language, was inescapably a matter of scientific culture as opposed to universalizable abstraction, and was tantamount to an imposition of normalization on the natural world by way of graphical instruments. As de Chadarevian argues: "Thus, the universality of the inscriptions and the phenomena they represented were the outcome of the negotiation of common standards which disciplined experimental practice and normalized experience. They were 'natural' only for the participants of this same scientific culture" (1993, 290).

EXTRAMURAL PHYSIOLOGY: THE *FUSIL PHOTOGRAPHIQUE*
AND THE PHYSIOLOGICAL STATION

The physiological laboratories of Magendie and Bernard were, for Marey, historically tainted sites due to their reliance on vivisection. The increasingly portable instruments and devices Marey was developing freed him to pursue physiological investigations in the wild, as it were. The portable sphygmograph he co-developed in 1859 was an early instance, but later devices included a series of portable odographs fed by rubber tubes attached to the shoes of a human runner to track step count, and to a horse's hoof to track gait, as documented in *La Machine animale* (1873). The great advantage of Marey's famous *fusil photographique* was that it was a highly portable means of producing a rapid image sequence consistent with the graphic method. Unlike Muybridge's multicamera setup, a photographic sequence could be produced in the field, with a single camera onto a single plate. Pierre Janssen's original "astronomical revolver" on which it was based was designed to chronicle the successive positions of the planet Venus in the night sky, and combined photographic apparatus with a telescope to produce seventeen images on a hand-cranked rotating single glass plate. In 1876 Janssen had already seen the potential benefits for other applications of his invention, including the physiology of movement, and in Marey's book eight years later on the development of the graphic method and its use in photography (*Développement de la méthode graphique par l'emploi de la photographie*, 1884) he quotes Janssen approvingly:

> The characteristic of the revolver is that it affords an automatic means of taking a series of photographs of the most varied and rapid phenomena in a sequence as rapid as may be desired, and thus opens up for investigation some of the most interesting problems in the physiology and mechanics of walking, flying and various other animal movements. (Janssen, quoted in Rabinbach 1992, 105)

Marey wrote to Janssen to ask for the precise specifications of his device, but found that Janssen's revolver was suited to the slow procession of the planets but unsuited to the rapid movement of animals, and any shaking would distort the image. Marey therefore had to adapt the revolver into a more portable version that was far faster in exposure time and rotation of the plate. With its long barrel and lens, and a magazine with a rotating disc mechanism, Marey modeled it on a hunting rifle, hence the *fusil photographique*. Marey's version could take twelve photographs per second onto the same glass plate, with an exposure time of 1/720th of a second. No longer limited to having to attach instruments, rubber tubes, or switches directly onto the animals, Marey was now free to trace the flight of a bird in the sky, or a quadruped running across the landscape. Furthermore, at least theoretically, it provided "a measure of duration" (in Rabinbach 1992, 106). A significant problem with Muybridge's photographic sequence was the lack of temporal consistency. Although the successive movements of an animal could be shown, Marey and other scientists saw the method as prone to inaccuracy, as there was no consistent perspective, and the physical properties of the wires and inconsistencies with the setup meant that the timing of the photographic sequence was irregular. Essentially, as Braun puts it, "Muybridge had failed to represent the trajectory of the movement" (1994, 53). Writing many years later, Marey reluctantly acknowledged Muybridge's role in the development of chronophotography, but with this large caveat: "For the velocity of the horse not being quite uniform, the equidistant wires were not reached at equal intervals of time. Besides, the wire was more or less stretched before rupture took place. From these causes there was a certain inequality in the rates of succession which Muybridge did not succeed in satisfactorily overcoming by letting off the shutters independently of the horse's motion" (Marey 1902, 319n). However visually impressive Muybridge's motion sequences would become, and how it captured the public's imagination (see, e.g., Phippen 2016), it would never become truly scientific until the dimension of time was systematically incorporated. Marey's *fusil* was designed to overcome the limitations of Muybridge's solution in this respect: "a single point of view, and equal as well as verifiable intervals of time; continuous registration on a single plate instead of instantaneous images on multiple plates," explains Braun (1994, 61). In practice, the images obtained were often indistinct, and because the *fusil* was handheld, the distance traveled by the photographic subject could not be measured easily. A further development of the *fusil* permitted multiple successive exposures on the same glass plate, leading to a striking composite image of a single human body leaving a trace of its movement of torso and limbs as it passes by. In keeping with the graphic method, and as part of the extensive genealogy of graphic inscriptors from Thomas Young onward, Marey first termed his means of decomposing movement "photo-chronographie," but then settled on "chronophotographie" (Rabinbach 1992, 103).

The many adaptations of existing instruments, graphic inscriptors, and portable measuring devices over the years led to perhaps the largest apparatus Marey was involved with so far, the Station Physiologique. The first scientific site of its kind, it would include a host of innovative equipment and measurement devices spread throughout. Writing first for the popular French scientific journal *La Nature* in 1883, and then for *Revue Scientifique* in 1894, Marey explained the rationale for instituting his Physiology Station outside the walls of the usual scientific laboratory. In *La Nature* he had highlighted the healthy character of this new research site, along with his broader aim to reunite physiology, "daughter of anatomy," as he put it (1896, 393), with zoology through the detailed observation in situ of animal, but also human, locomotion:

> The physiologists of today use new methods and instruments of precision which enable them to study the phenomena of life with an exactitude formerly obtained only by physicists. This apparatus, first intended to be used in vivisection, is undergoing a gradual change and is tending to become applicable for use upon animals and to man himself in perfect health and in the exercise of their normal functions. (392)[7]

Where better to study the exercise of the normal functions of animal and human locomotion than out in the open air, away from the laboratory and its associations with dissection, and with the necessary apparatus all available on-site? A bird's-eye view of the station would show two large concentric rings around the perimeter and an inner circle, collectively appearing almost like an eye. The outer track of five hundred meters was for human runners, and the inner track for horses. Telegraph poles were distributed at the perimeter of the outer track at regular distances with electric relays. As Braun (1994) explains, the main building or "chalet," built in December 1882, was large enough to accommodate Marey's on-site laboratory along with an increasing number of instruments and measurement devices (see Figure 14). The electrical relays around the perimeter were connected to an inscribing odograph housed in the chalet so that the speed of a runner could be plotted accurately, without having to wear special rubber tubing or shoes with pneumatic switches. In terms of photographic equipment it would come as no surprise that, after the experiments with his *fusil photographique,* Marey sought an alternative to Muybridge's fixed series of trackside cameras. On a short section of railway track extended perpendicularly from the perimeter toward the center, a small wagon (*la chambre noir roulante,*

FIGURE 14. *Étienne-Jules Marey at his desk with measuring instruments, around 1900. Source: Wikimedia Commons.*

or "rolling camera obscura") was placed that housed a single-plate camera. Perhaps more like a *canon photographique* than a *fusil,* the images could be larger, more distinct, and more stable than the handheld *fusil* and "produced rapid exposures at regular intervals of time" (Marey, in Rabinbach 1992, 106) by means of a rotating shutter mechanism onto a fixed glass plate. This camera pointed toward an area of the perimeter with a dark backdrop, an *écran noir* (black screen) against which the moving subjects, wearing white and illuminated by the sun, could be photographed more clearly. The rotating shutter, with regularly interspersed windows for light, therefore recorded successive phases of the subject's movements at precise intervals onto the fixed photographic plate, resulting in a single compound image. In July 1882 the first photographs with that system of a runner were published in *La Nature,* and soon afterward Marey added a luminescent chronometric dial to the front of the screen, so that images of moving bodies in space could be more precisely analyzed in time. This interface was clocklike, and so the spacing of the dials clearly marked time intervals that were ingrained in the photographic image itself, the better to aid analysis. "The chronometer signified the complete integration of time into the study of motion," states Rabinbach (1992, 107). In the summer of 1883, regularly spaced black-and-white colored markers were placed in view of the camera to further aid visual analysis. Each of the two chronophotographic systems, the rotating plate and the fixed plate, had their distinct advantages. "When it is wished to record the movements as a whole, it is necessary to use chronophotography on a movable plate, for it gives a series of entire images of the subject in action," Marey explains (1896, 399). For example, to analyze human walking in this way, the successive bodily attitudes in taking steps are shown discretely by means of successive images on the same photographic plate. The velocity, as well as the measure of displacement of muscle and limbs, can be derived by the further labor of isolating each image and placing them in order, and then making a series of manual ink tracings of the outline of the limb, plus bone or muscle placement. The result is a geometrical abstraction of the movement involved, in keeping with the graphic method, but the movement analysis is a long and laborious process. The fixed plate, on the other hand, involves a single long exposure resulting in "a photographic outline of a movement," explains Marey, where "the images of a moving man or animal are reduced to a few brilliant points and lines, but this is generally sufficient to mark the action of the limbs in the different gaits" (399). In other words, the geometrical abstraction of movement is already visible in the photograph, written in through the long exposure. This method Marey found worked not only for gross locomotor phenomena, as in walking or running, but also for the scientific study of smaller-scale motor functions such as chewing, respiration, or partial limb movements. For activities that involved particular techniques of movement, such as a gymnast leaping or a blacksmith hammering at an anvil, the fixed-plate multiple-exposure image accentuated this repetition, drawing the observer's attention to concentrations of bodily movement. His collaborator, the engineer Charles Fremont, had conducted such a study at the station, as Marey describes:

> A distinguished engineer, M. Frémont, has just made at the physiological station a study of the work performed in hammering iron upon an anvil. On a movable film M. Frémont took the series of attitudes assumed by the blacksmith in the successive movements given to the hammer. By chrono-photography on a fixed plate there

was shown the trajectory of the hammer and successive positions at definite instants, the intervals between which were measured by a chronographic dial. (402)

As with the rotating plate of the *fusil,* however, Marey found the limitations of the fixed-plate device frustrating at first. The long exposure times meant that it was well suited to fast movement of subjects across space, but the rapid deceleration of a hurdle jumper, for instance, or the arm movements of a blacksmith, resulted in rather blurred and indistinct compound images. "Problems of legibility linked to the overlapping, blurring, and superimposition of figures were due, in a sense, to the fact that there was *too much* detail in the photographic method," as Mary Ann Doane puts it (2002, 54, emphasis in original).

In 1883 Marey overcame this problem by means of what he termed "partial" or "geometric" chronophotography. For Marey "it was desirable to multiply the images while showing the whole body" (1902, 324) for the sake of clearer locomotion analysis. Were the photographic subject to wear silver stripes along the limbs or patches on body parts of interest, their movements would be literally highlighted, made more visibly outstanding against the dark background, as a result of the long compound exposure. Were the subject to wear a black suit that covered their entire body, including their head and face, the highlighted body parts would further stand out against the dark background and, because of the rapidly rotating shutter, the geometric movement patterns are all that are visibly captured. As Marey explained retrospectively:

> A man dressed completely in black, and consequently invisible upon the dead-black background, wears certain bright points and lines, strips of silver lace attached to his clothes along the axes of his limbs. When this man, so rigged, passes in front of the apparatus, photographs will result that will be accurate diagrams to scale, showing without confusion the postures of upper and lower arms, thighs and lower legs, and feet at each instant, as well as the oscillations of the head and of the hips. The method also allows the play of the joints to be studied. (Marey 1902, 323)

A photograph of the man dressed entirely in black, with his head covered entirely with black cloth, has been reprinted in Marey's own publi-

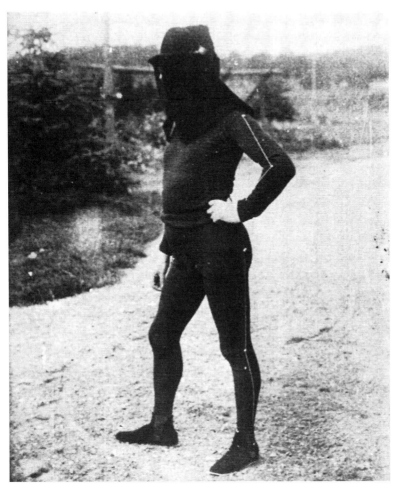

FIGURE 15. *Georges Demenÿ wearing the black suit and hat, with silver stripes, 1883. Source: Wikimedia Commons.*

cations including *Le Mouvement* (1893) and in the secondary literature (see Figure 15). The man underneath the suit is in fact Marey's assistant Georges Demenÿ. The "partial" or "geometric" form of chronophotography led to a strikingly high-contrast image sequence of luminescent skeletal body parts exhibiting movement sequences that ordinarily the eye could not follow. Rather than the confusing blur of rapid movements of before, the reduction of the body to a geometric pattern of lines across space could be more easily and accurately tracked, especially in phases of acceleration and deceleration. This stark difference can be seen by contrasting two chronophotographs from the same year, 1883. The blurred movement within the fixed plate appears in *Demenÿ Walking,* where an aggregate overexposed and indistinct image of Demenÿ walking across the frame makes it seem as if there is a dense crowd of ghostly figures blurrily spreading across the picture. Conversely, the geometric chronophotograph *Joinville Soldier Walking* (see Figure 16) using the black suit similarly involves the whole body moving across the frame, but the illuminated sequence of stripes clearly depict the trajectories of individual arms and legs, and "allows the play of the joints to be studied" (1901, 323)—far more effective for gait analysis. The subtraction of unnecessary detail, the way that more and more of the background and the body of the subject receded into the black background, meant the result was geometric, abstract, or, as Doane describes, "a series of chronophotographs consisting only of lines and curves in space" (2002, 54). The chronophotographic outcome therefore folded readily into Marey's more general "graphical method," even if the graphic method stressed continuous curves rather than the intermittent stutter of geometric shapes. Writing on the capture of human movement in his *Movement* (1895), Marey explores the implications further: "In geometrical photographs, thanks to the great number of images, the discontinuity of the phases almost entirely disappears, and the actual path followed by each point of the body can be represented almost as a geometrical curve" (145). However, a later chapter entitled "Applications of Chronophotography to Experimental Physiology" addresses the contrast between the continuous curves involved in the graphic method, in this case a cardiogram, with the successive images of the chronophotographic method. The graphic method, he now concedes, "only appeals to the initiated," and the appreciation of the significance of what the cardiogram reports actually results from numerous prior experiments and observations, and is therefore not immediately intuitive or self-evident. On the other hand, Marey continues, the chronophotographic method "is the direct examination of the movements . . . by a more subtle eye than ours, and one that is capable of grasping in a moment the sum total of the changes which take place." By offering an immediate comparison between a series of consecutive images, the method "affords an opportunity of observing every visible phase of the phenomenon" at a glance, yet inescapably "its record is one of intermittent indications, instead of the

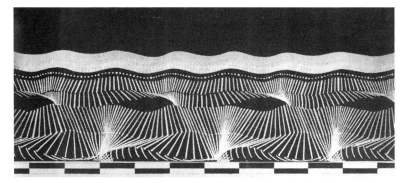

FIGURE 16. Joinville Soldier Walking, *1883. Source: Collège de France Archives.*
3PV65.

continuous record of a curve" (287). Muybridge's less scientific, yet aesthetically captivating, motion sequences had fixed animal locomotion
in space, then, if not in time. Even so, the spatial separation of the photographed figures meant that tracing the movement of figures from one
position to another was unsatisfactorily unscientific for Marey. Geometric chronophotography offered far greater and immediate legibility
for motion analysis, and according to Doane was "much more suited
to the representation of space than is the graphic method" (2002, 56).
Chronometric dials could still be incorporated into the background to
visibly slice time into regular and discrete instances, fixing these intermittent geometries across space. Nevertheless, the project of chasing
ever-smaller increments of the passage of time through the refining of
the chronophotographic process meant that, as a continuation of the
graphic method, he was constantly in search of "a technical apparatus
that could make visible minute changes over time—'les infiniment pe-
tits du temps'" (Rabinbach 1992, 107). Later experiments with moveable
film and elementary spooling mechanisms with bobbins in 1887 would
be another logical step beyond the fixed photographic plate crowded
with bodily movements, and Doane's chief interest in Marey is in how
such equipmental innovations cumulatively approach cinematography.
    Marey had recruited Demenÿ in 1881, and his assistant's first task was
to oversee the completion of the Physiological Station. Marey, Demenÿ,
and a municipal councilor named Hérédia cannily pitched for it to be
funded through the city of Paris and the Ministry of Public Education,
since its findings would have practical applicability, improving the

health and bodies of the nation. In fact, Marey and Demenÿ spent much effort in popularizing the project for the general public and justifying it to potential funding sources. In an 1883 article in *La Nature,* translated and published in a report for the Smithsonian Institution, he took the opportunity to explain what the station complex was designed to do, and speculated about what it could do:

> The physician may seek there new means for the diagnosis of certain maladies and for investigating the effects of their treatment; the soldier may study there the proper regulation of marching so as to diminish fatigue and use to greater advantage the bodily forces; the educator of youth may learn how to logically direct gymnastic exercises; the artist, how to represent more truthfully the scenes that he wishes to depict; the agriculturalist, how to use to the best advantage the strength of animals; the artisan, how to more quickly acquire the skill necessary for his professional labors. (1896, 407)

For the purposes of funding, among its stated aims was "research into the conditions for the best use of muscular force in men and animals," and tipping his hat to the minister of war, "to traverse the greatest distance with the least expenditure of energy" for soldiers (in Braun 1994, 71). Another was to produce an improved and more efficient method of gymnastic instruction for the teachers of Paris, leading to updated manuals, as "there was a real lacuna to fill in this area, for right now absolutely no rational teaching for this science exists" (in Braun 1994, 71). Another justification was an optimization of movement and gesture for use in the factory, as stated in his book *Movement*: "The same method would lend itself to the teaching of movements necessary in various vocations. It would show how the stroke of a skilled blacksmith differed from that of an apprentice. It would be the same for all manual acts." (Marey 1895, 139).

Along with Muybridge's later photographic work, the practical benefit of the new graphic physiology would have a profound impact upon nascent ergonomics, and be the starting point for early twentieth-century work on labor and fatigue by Taylor, Gilbreth, and others. "Why this attraction to gymnastics and athletic movement?" questions Dagognet. "Why the orientation of neuro-motor physiology toward motor activity?" (1992, 168). Demenÿ saw the sudden popularity of gymnastic socie-

FIGURE 17. *The Physiology Station at Parc des Princes. View from the track, from Marey (1883, 228). Illustration reprinted with permission from CFC (Centre Français D'Exploitation du Droit de Copie).*

ties and an emphasis on physical strength that could be corralled into martial interests as emerging in the wake of France's humiliating defeat in the Franco–Prussian war in 1870. Collectively, Marey and Demenÿ's justifications were convincing enough that funding was granted in 1881, and increased in 1882. After some difficulties siting it, an agreement was made at the Parc des Princes, a public park in Paris's sixteenth arrondissement (see Figures 17 and 18). Incidentally, in 1883 Muybridge similarly campaigned for, and instigated, an ambitious photographic research program involving the building of an external testing station at the Philadelphia zoological gardens, underwritten by the University of Pennsylvania (Muybridge 1955, Taft's introduction). By June 1884 he was photographing on the grounds of the veterinary department. His collaborator Thomas Eakins even photographed human movement with the fixed plate "Marey-wheel camera," according to Brookman (2010, 92), but Muybridge mostly defaulted to a variation of his earlier fixed battery of electromagnetically operated cameras to shoot their subjects against a marked grid background.

"In Marey's images the body became a trace on a glass surface, a

FIGURE 18. *The handwritten caption reads: "Bird's-eye view of the principal buildings of the Station." Photograph: Collège de France, Fonds Marey (13 Fi 5, V, 1), with kind permission.*

trajectory of decomposed movement," as Rabinbach (1992, 108) puts it. Although firmly situated in the scientific process, and made invaluable for a new epoch of experimental physiology, there would be far broader and unforeseen cultural impacts of Marey's work and Demenÿ's continuation of it at the National School of Gymnastics and Fencing at Joinville after Marey's death (see Figure 19). Obviously, there is the equipmental refinement of the means of capturing continuous motion that led to a plaque at the house where he lived in Paris, near the Physiological Station, commemorating that he "had taken part in the invention of cinematography" along with the Lumière brothers (Dagognet 1992, 163). Apart from his contributions to a nascent cinema, the impact of the newer medium of chronophotography on the older medium of painting was profound, to some extent echoing Muybridge's crossover. Earlier I mentioned the visual debt that Duchamp owed to Muybridge's photographic sequence *Woman Walking Downstairs*, but Duchamp also acknowledged the influence of Marey's geometric sequences on his *Nude Descending a Staircase, No. 2,* especially the dotted lines around the hip and leg. Duchamp's juxtaposition of forms and the elongation of parallel instances of those forms were present in earlier paintings, with the mapped-out rotation of the handle of his schematic *Coffee Mill* (1911), and *Sad Young Man on a Train* (1912) having the most stylistic resemblance of a bodily pose. In a published interview, Duchamp refers to Marey's monograph *Le Mouvement* (1894) as one of the sources for this artistic direction: "In one of Marey's books I saw an illustration of how he indicated the fencers or galloping horses with a system of dots delimiting the different movements" (in Braun 1994, 287–91). Stylistically, then, Marey's photographs presented him with modifications of Cubism, thinks Braun: "In more general terms, chronophotography introduced Duchamp to the idea of mechanization as an alternative to sensible beauty" (1994, 291). In the wake of Marey, if modernist artists were retreating from the immediately perceptible in order to respond aesthetically to the machine-made image, Doane ultimately sees the failure to fix time sufficiently within chronophotography, and therefore the failure to fulfil Marey's own aspirations of the graphic method, as in fact the basis for aesthetic inspiration: "It is significant that the limit or failure of Marey's scientific endeavor—the blurred image—was subsequently taken up by modernism (especially Italian Futurism) as evidence that the perfect representation of time (particularly its more 'modern' aspects of speed and dynamism) was precisely illegibility (nondifferentiation)" (2002, 60).

Meanwhile, a year after Muybridge and Marey's deaths, a paper on improved methods for high-speed chronophotography was published by Alfred Gradenwitz, who documented how Lucien Bull's version of the apparatus removed the fast-rotating shutter and instead synchronized the movement of film with illumination from timed electric sparks. Whereas Marey was able eventually to capture the movement of insects' wings at 1/25,000th of a second, Bull's method of electro-illumination offered a pathway up to two millionth of a second (Gradenwitz 1905, 51). A further step in high-speed flash chronophotography was made by MIT professor Harold Edgerton, and Albanian photographer Gjon Mili created strikingly clear and luminescent images for *Life* magazine in the 1940s with this equipment. Mili's 1941 photograph of U.S. pentathlon champion John Borican has a clarity in the multiply exposed

FIGURE 19. *Georges Demenÿ,* Fencer, *1906. Demenÿ established the National School of Gymnastics and Fencing at Joinville. Photograph: public domain, courtesy of the Metropolitan Museum of Art, New York.*

trace of movement in throwing a javelin that Marey could only have dreamed of, for example, although their purpose was not scientific but, like Muybridge's photographs, an aesthetic enterprise.

Analysis and Synthesis of the Biomechanics of Movement

Machine-made images were indispensable for the analysis of bodies in motion, then. But Marey made an additional step after analysis toward the synthesis of motion. In his wish to model physiology more closely on the theoretical mechanics of physics and chemistry, to make recordings without human intervention, Marey also realized and materialized the reverse of this process. From the machine-made images, with collaborators he actually *made* machines that furthered the reconstruction of movement after its initial decomposition. Braun identifies this as two complementary methodological strategies resulting from Marey's use of mechanics as an explanatory model: "The first was to submit the phenomena being studied to measurement by machines. The second was to construct mechanical models that would simulate the phenomena under investigation" (1994, 15). Since the physicists and, to a lesser extent, chemists, would construct mechanical models to verify experimental results, Marey would do the same for the results given by his graphic inscriptors. To some extent, this was in keeping with his German physiologist antecedents, especially Wilhelm Wundt. If, as previously discussed, vivisection as the violent disassembly into constituent parts was "the first step on the ladder toward knowledge of an organism's interior world," as Dierig (2003, 128) has it, the interior processes thus revealed could only be fully understood if the experimenter could somehow re-create "the same conditions and the same phenomena" outside the organism, according to Wundt in his *Lehrbuch der Physiologie des Menschen* (1865, Textbook of human physiology). This next step, the "artificial reproduction of nature's phenomena," continues Wundt, must "always be the last step of experimentation" (in Dierig 2003, 128). For any such scientist, the resulting mechanical models were never taken to be a true or even representative picture of reality, but rather "a demonstration that phenomena could in principle be represented by mechanisms and that such constructions rendered them intelligible, since the conditions of their existence could be reassembled," as Braun puts it, and constituted "a mode of synthesis" (1994, 22). What the cardiogram had measured early in Marey's career was subsequently reconstructed through artificial means, for example, and so he constructed a model circulatory system using rubber tubes and chambers in 1857 (19). He called such models a "schematic apparatus" *(appareil schématique)* in an 1869 essay on the mechanisms involved in insect flight. Indeed, the most astonishing example follows from his interest in the movement of birds and the mechanisms of flight. Remembering that Marey had wished to depart from the unnecessarily destructive methods of vivisection routinely employed by his forebears, his apparatus over time became ever more abstract, complex, and physically distanced from the animal itself. The creation and construction of mechanical models could therefore be regarded as an antivivisection. At first he harnessed insects and birds closely to long horizontal sections of glass plate, the movement of the wings inscribing directly according to the graphic method, but Marey found that the movement recorded was uncharacteristic. More loosely tethered and therefore allowing more naturalistic movement, in 1869 the bird harness included tambours, another of Marey's inventions, which independently recorded both the vertical and horizontal movement of each wing based on the transmission of air pressure, allowing direct inscription of the wings' full motion through rubber tubing. The *fusil photographique* was clearly a more liberating apparatus, able to chronophotographically capture untethered movement of birds in the wild. From 1868, Marey's fascination with mechanical flight, and the possibility of the invention of actual flying machines, led to a series of mechanical insects with the assistance of Victor Tatin. With Tatin's increasing involvement, from around 1870 they constructed mechanical birds with moving wings, plugged into tubes and tambours, together simulating the patterns of winged movement already recorded through the graphic method. One of the models designed by Tatin in 1879 using compressed air to power two propellers even flew for a few seconds. In 1886–87 he crafted "the first of several model 'syntheses' of flight in the form of wax or bronze sculptures of the successive phases of a gull's movements," as Rabinbach (1992, 108) describes, three-dimensional chronophotographic realizations of the trajectory of animal movement, redolent of the futurist sculpture of Umberto Boccioni.

Such syntheses were intended to model movement according to multiple physical dimensions, and there is an imaginative step in transcribing the data of movement into expressions through other media. But the synthesis of movement perhaps most enduringly occurred in returning to the two dimensions of the photographic plate. Muybridge's demonstration of his motion sequences in the zoetrope Rabinbach considers as "the first dramatic example of the photographic synthesis of movement" (1992, 102). Accordingly, the synthesis involved in the perception of artificial movement based on the combination of Marey's *fusil photographique,* the rotating plate camera that recorded sequences of bodies in movement in regular and discrete time instances, and the zoetrope, which allowed playback, slowdown, and therefore analysis, is the most apt example here. By adapting and enhancing the resulting chronophotographic sequence, and then placing it inside a zoetrope, it reconstructs that movement before the retina, performing a perceptual synthesis for the viewer. The familiar psychological trick of the persistence of vision, "whereby the brain retains an image cast upon the retina for an instant after the object viewed is removed or changed" (Braun 1994, 28) was exploited by the zoetrope and other nineteenth-century "philosophical toys," like the phenakistoscope and the thaumatrope, and that, as we saw with Muybridge, also operated with projection machines like the magic lantern and, courtesy of Marey, later the cinematograph. Although they exploited common visual illusions, the term "phenakistoscope" literally deriving from the Greek word for "trickery" or "cheating," such toys remained philosophical insofar as they were active demonstrations of how our mechanisms of perception elicited the illusion of continuous motion. For example, the lesser-known physiologist Adolphe-Moïse Bloch noted how the visual sense functioned like the magic lantern that so captured Muybridge's audiences, one that "constantly fused discontinuous reality and provided a sense of movement," in the words of Canales (2009, 79). This was an "unconscious synthesis," as Bloch wrote in the *Revue Scientifique* of May 1887:

> It is the instinctive education furnished by visual persistence that makes the postures of passersby in instant photographs seem so bizarre. In our sensorium, every movement is tied to that which precedes it and that which will follow. We do not see the galloping horse, as it is, but as it was and as it will be in a fraction of a second. So it is this continuous, *unconscious synthesis* that takes away the possibility of accepting as real the truth itself of things, as when it is shown to us by a photograph. (in Canales 2009, 79, emphasis added)

Chronophotography was a continuation of the graphic method, and what is true of the graphic method, chronophotography, and later cinema as well for Marey, is his fascination with these as mechanisms of legibility, representation, and even storage: "The cinematograph would later be characterized by the same ability to receive, retain and rerender" (Dagognet 1992, 38). Or, in comparable terms, both graphic method and photographic method involved "an interceptor to capture the movement, a transmitter to relay it, and an inscriptor to make it at once visible and permanent" as Braun (1994, 61) puts it.

In his 1898 article "Analysis of the Horse's Movements by Chronophotography" in *La Nature,* translated and reprinted in a supplement to *Scientific American* that year, Marey reiterates how the chronophotographic method helps him "to study the anatomy and physiology of the locomotive apparatus comparatively" (1898, 18924). This was not based on any naïve assumptions about the photographic images themselves, even if the images could now be captured outside the laboratory and "in the wild," as it were. The display of successive photographic images in a sequence would only replicate what Muybridge and others had already achieved—namely, "a study upon the exterior of the horse by means of chronophotography" (1892²). The unconscious synthesis that produces the simulation of animal movement is not enough, and the further step of graphic inscription is necessary:

> The photographic images that are projected successively at very short intervals become blended upon our retina into a continuous sensation which strikingly *reproduces the appearance of the motion itself.* But such images scarcely answer for the physiologist, who perceives therein nothing more than what an observation of nature would show him, that is to say, actions so rapid and complex that he is unable to seize the phases thereof. (1898, 1892²; emphasis added)

Impressed as Marey was at first on encountering Muybridge's technique in 1878, then, it offers merely the "appearance" of movement, since a "series of images contains all the elements necessary for the perfect knowledge of the motion of animals; but to this effect, it is necessary to unite the images in a general figure with their relative positions in space," says Marey, as "a comparison between separate images is too difficult" (1892²). Placing successive images in a single horizontal strip, as Muybridge had done, is a step toward the synthesis of a general figure of equine locomotion. In the 1898 publication he offers three consecutive vertical strips, each one showing a different gait, without and then with rider for a more synoptic comparison of general figures of movement. Nonetheless, Marey makes it absolutely clear that the analysis of the chronophotographic sequence was the start, and not the end, of the process. Whereas for Muybridge augmentation of the photographs through retouching was a necessity for the purpose of clarity and to foster wider dissemination, for Marey the chronophotographic sequence was merely the intermediate step to the diagramming of the biomechanical movement beneath the skin, and only then could one derive a "true geometrical diagram of the motion under study" (1892²). To more effectively diagram movement and how the forces are transmitted dynamically across the musculoskeletal frame, the marks and traces inscribed upon the photographic image would be more significant for Marey than the image itself. Marey's use of figures reproduced in the 1898 text mirrors this movement from the geometric chronophotography to the pure geometrical diagram of movement, since of the three figures only one involves photography, the other two are line drawings derived from the photographic sequence that superimposes skeletal and musculoskeletal diagrammatic supplements. Individually, onto photographs of the movement sequence are traced lines and curves so that "it is possible, even, to combine the form of the bones and muscles with the diagram of the external forms, and thus obtain a knowledge of the hidden mechanisms of which the motions of the animal are merely an external manifestation" (18922).

This amounts to the realization that, despite the significant advances in the techniques of chronophotography that Marey, his collaborators, and successors will make to significantly reduce the temporal stutters and shuttered gaps in the mechanical capture of movement, the hands-on craft of tracing the complex biomechanical process of musculoskeletal movement with pen and ink onto the photographic frame was far more significant for further physiological analysis. Moreover, the value of the chronophotographic method at the Station Physiologique was in part the liberation from vivisection, as Marey had justified earlier. Regrettably, for the process of producing some of these diagrams, horses that were first chronophotographed were then euthanized, the better to comprehend and then depict movement arcs of bone and muscle groupings. For example, the stallion Tigris suffered the same fate as other horses retired from service and executed, but in his case he was taken to the Physiological Station to be euthanized and then dissected, the better to map the underlying musculoskeletal topography. Under the skin, behind the inscribed diagrams of movement, the phases of elongation of particular muscle groups and antagonistic reactions by corresponding groups are revealed. The scapulohumeral muscles around the shoulders, for example, help determine the shifting of weight across the skeletal frame according to each movement phase. Or how, in Marey's words, "the osseous projections of whose relief is outlined under the skin fall well into their places" (18923) revealed far more than the position of the limbs at any one time. Marey's graphic method had consequently become revitalized through the chronophotographic technique, but the actual photographs themselves receded from visibility once the biomechanical diagrams—the distribution of forces and the effects on clustered mechanisms of bone and muscle beneath the skin—were revealed.

From Capturing Movement to Measuring Work:
Fatigue and Efficiency

At various points throughout this history, an awareness of fatigue appears as a potential object of study, and as a practical application for the new chronophotographic means of measurement. As a prelude to the proceeding chapter, whose overriding topic involves the measurement of human fatigue in line with the graphic method, some final observations about the legacy of Muybridge and Marey is offered. One practical application of chronophotography based on Muybridge's pioneering approach was his transition to the observation of the movements of workers, which impressed Marey. In *La Machine animale* (1873), Marey's interest in the measurement of animal movement prompted consideration of efficiency, and so the arrival of chronophotography became another technique for the calculation of "the mechanical work expended in different movements," as Marey described in a letter to Demenÿ in 1881 (in Rabinbach 1992, 116). Marey's previously mentioned paper for the International Congress of Hygiene in 1884, "The Motive Power of the Human Body," continued this line of inquiry. As covered in more detail in the next chapter, the adaptations of familiar devices such as the myograph by Helmholtz between 1848 and 1852, and by Kronecker in the 1870s and Mosso in the 1890s, among others, began to establish the measurement of certain muscles and therefore fatigue in the laboratory. The familiar curves and arcs inscribed as outputs were of course consistent with the graphic method. The period in which Marey was experimentally active also offered significant leaps in the foundational science of fatigue, and there is certainly crossover in terms of methodologies, instruments, and empirical focus on bodily movement. In 1867 Marey had reproduced some of Helmholtz's experiments with the myograph, measuring the muscular movements of frogs' legs, which he detailed in chapter 4 of *La Machine animale*. Marey's adaptation of the myograph is illustrated and explained: "It shows the disturbance of the muscle by means of a curve which readily allows us to study its phases" (1874, 31–32). But of course the muscles and bodies within the laboratory myograph are fixed. Either a frog is decerebrated and its entire body is fixed to a cork board by long metal pins, or the frog's calf muscle is removed entirely and suspended between metal grips, one attached to a stylus that inscribes upon the rotating drum. The arrival of the chronophotographic method offered an alternative to this fixity. In his address to the Association Française pour l'Avancement des Sciences in 1886, "Étude de la locomotion animale par la chronophotographie" (Study of the movement of animals by chronophotography), Marey directly acknowledges the link between the practical application of the measurement of fatigue and the possibility of more efficient patterns of movement through the new apparatus: "As we regulate the use of machines in order to obtain a useful result with the least exertion of work, so man can regulate his movements . . . with the least fatigue possible," he declares (in Rabinbach 1992, 116). This would result in another form of graphic notation birthed from physiology that presaged the motion studies of Frank and Lillian Gilbreth early in the twentieth century, who themselves will use adaptations of chronophotographic techniques to trace movement in space such as their "cyclogram," as will be shown. In the wake of Muybridge's photography, Marey had more fully established how a physiology based first on the graphic method, and then on chronophotography, could help determine the most efficient movements of the body, and serve as an early articulation of the idea of "ergonomics," the science of efficient movement. Both the odograph and chronophotograph, central instruments within the Station Physiologique, were part of this practical application of calculating expended work from movements, and hence also precursors to ergonomics and the scientific management of Taylorism. As Rabinbach summarizes: "Chronophotography made possible a science of fatigue and a rationalization of the body's movements—an economy of energy that led to a distinctly European science of work" (1992, 116). Toward the end of his career, the first application of chronophotography to the study of physical labor was realized by the engineer and longtime Marey collaborator Charles Fremont in Marey's laboratory in 1894, tasked with investigating optimum work performance. His studies were published in 1895 in a popular Paris monthly publication *Le Monde moderne*. Another example of what can be productively learnt in the comparison between the scientific observation of movement and the ways it is typically depicted in the visual arts, Fremont showed his chronophotographs in counterpoint to previous paintings and sketches of workers doing the same actions, such as blacksmiths with a red-hot anvil and a forge in eighteenth-century sketches by Gaspard Mongé, and even engravings from Denis Diderot and Jean le Rond D'Alembert's *Encyclopédie*. Dagognet is far more critical of this more practical direction. "If he caught a glimpse of the importance of the corporeal, its rhythms and psychomotor powers, through dismantling gesture, he hastened standardization, surveillance, production goals and alienation," he says (1992, 185). The fact that Marey was far from alone in his scientific application of the study of movement in the workplace, as this aspect of his research continued through figures such as Amar, Mosso, Taylor, and the Gilbreths, forms the trajectory of the following chapter.

One last observation, to return full circle from the earlier description of Turner's painting of the fast-moving train by the reviewer in the *Times* of 1844 as "realities seized upon a moment's glance." The last words and thoughts go to Henri Bergson. At one point Bergson refers to Marey's "cinematographical mechanism" (1911, 332) within a longer explanation of what he terms the "image." By "image" he does not mean the usual fixed illustration or photographic capture of movement, but rather a technique outside of our everyday consciousness for artificially halting the usual flows of movement in time, to conceptualize what lies behind the sensible qualities of matter and duration. He addresses this chronophotographic way of thinking in *Matter and Memory* ([1896] 1999) and *Creative Evolution* ([1907] 1911). In one passage in the former, Bergson somewhat mirrors the *Times* reviewer's description, putting this in equally poetic terms as "the glance which falls at any moment on the things about us," and which seems to encompass the analysis and synthesis of movement, its perception and reproduction, and therefore the consciousness of continuity, all at once:

> In just the same way the multitudinous successive positions of a runner are contracted into a single symbolic attitude which our eyes perceive, which art reproduces, and which becomes for us all the image of men running. The glance which falls at any moment on the things about us only takes in the effects of a multiplicity of inner repetitions and evolutions, effects which are . . . discontinuous, and into which we bring back continuity by the relative movements that we attribute to "objects" in space. (Bergson [1896] 1999, 277)

Alas, would that we had the time and the space to pursue this in more detail.

CHAPTER 5

Fatigue

Jules Amar, Angelo Mosso, and Physiological Observations of Industrial Labor, 1891–1947

In the early years of the twentieth century, two eminent professors left their offices, rolled up their sleeves, and began work in nearby factories. Max Weber, a professor of political economy and one of the founders of sociology as a discipline, traveled to the small German town of Oerlinghausen in September 1908, and again several months later, to engage in preliminary fieldwork in the family-owned textile factory of Carl Weber & Co. The other, a future Nobel-winning scientist at Oxford named Charles Scott Sherrington, rode the sixty miles to Birmingham on his bicycle to do shifts on the production line of the Vickers-Maxim munitions factory in 1915. What was the impetus behind the extension of the field and the laboratory into industrial settings? Although the researchers were clearly from seemingly incommensurate disciplines, with very distinct methodologies and approaches, both were responding to new bureaucratic exigencies that sought to observe, and then to measure, efficiencies with human labor and the cost of fatigue. In the case of the munitions factory, reducing faulty output because of long shifts and fatigue, the reasoning went, might save accidental deaths. But there is another, more conceptually entangled answer that involves scientific developments in Germany, France, and Italy in the last decades of the nineteenth century, and the beginning of the twentieth.

We have already seen (in chapter 2) how the legacy of psychophysics, the Leipzig school starting with Wilhelm Wundt, Ernst Weber, and Gustav Fechner, was based on the identification and precise measurement of bodily sensations such as touch, pain, and color. From Weber's experiments on touch in *Der Tastsinn* (1834) onward, and Fechner's *Elemente der Psychophysik* (1860), the crafting of precise instruments for the laboratory went hand in hand with a new means of mapping comparative sensitivities across, and between, the bodies of experimental subjects through the introduction of the "just-noticeable difference"

(JND), also known as the Weber–Fechner Law. If the JND functioned as a means to establish normative thresholds of sensation for touch, pain, and the perception of color across populations, any "difference" to be noticed was ineluctably predicated upon an individual unit of experience, and therefore an introspective and subjectively variable element. Meanwhile, at the opposite end of the spectrum, more abstract theories concerning the laws of conservation of energy between animals and machines, most compellingly realized by Herman von Helmholtz around 1862, remained attractive in their application to the rapidly industrializing economies of Western Europe. Caught between the increasingly rationalized organizational bureaucracies that emphasized the logic of productivity and that considered labor as an abstract force on the one hand, and the physiological limits of repetitive movements of the worker's body over long shifts on the other, this is the story of industrial fatigue within human subjects and the need first to measure and then graphically depict it in terms of lines and curves.

This happens to be a story of two curves, the "work curve" ("Die Arbeitscurve," 1902) of Emil Kraepelin, and the "curve of fatigue" of Hugo Kronecker (1871), later reinvestigated by Angelo Mosso (*La Fatica,* 1891). The curves were inscribed on paper by means of specially constructed physiological instruments, part of a flourishing industry in itself at that time. Although this chapter opens with Sherrington's visit in 1915, the formation of the initial physiological problematic of fatigue, and the modification of measuring apparatus, takes shape at first through the observation of repeated contraction and expansion of the isolated muscles of animals—namely, Helmholtz's experiments on frogs in Leipzig between 1848 and 1852, and Kronecker's modifications of those frog experiments in Leipzig in 1873. Mosso in Turin had designed his ergograph in 1890 to measure muscular fatigue on human subjects, the research culminating in the publication *La Fatica* one year later, first translated into English in 1904. The nervous system of the industrial body was being charted, its movements, frictions, and tolerances examined, and the poor diets and the long hours of the working day were being factored in. Various social and political processes were to make "fatigue" a visible, and therefore a measurable, problem that could be rendered through graphical means into lines and curves. Starting with the extramural journeys of Max Weber and Charles Sherrington, and later the establishment of Britain's Industrial Fatigue Research

Board in 1919 and Lawrence J. Henderson's Harvard Fatigue Laboratory in 1927, this was a unique period in the swirling accumulation of ideas around economy, society, and human physiology in industrial settings.

The broad narrative of this chapter is set out in four parts. First, the section "Physiologists, Sociologists, and Factory Workers" considers the conditions that precipitated Sherrington's factory visit and his work with the Industrial Fatigue Research Board. Second, the section "Curves and Lines" highlights some of the developments within physiology that led to measuring instruments, from Helmholtz's frog-drawing machine to Mosso's ergograph, and the imperative to diagram the energetic work undertaken through a graph of movement over time. The development of Mosso's ergograph is seen in relation to similar precursor instruments such as Jastrow's aesthesiometer and Marey's sphygmograph, although the ergograph is able to accommodate the influence of environmental factors upon human fatigue. Third, the section "Movement and Fatigue" sees how Mosso's experimental findings were taken up by Jules Amar, involving Étienne-Jules Marey's more cinematographic physiology of human and animal locomotion. Amar's *Le Moteur humain* (1914) compiled extensive observations and data on the forces, movements, and thermodynamic processes involved in labor, yet remained influenced by Helmholtz's coinage of the term *Arbeitskraft* (labor power). Fourth, "The Age of Measurement" shows how Amar was a contemporary of the Americans Frederick Wilmslow Taylor, and Frank and Lillian Gilbreth. The Gilbreths' motion studies from 1907 stressed efficiency in movement through systematic observation in the laboratory, starting with brick-layers and then, echoing Sherrington's extramural physiology, munitions manufacture in World War I, but also brought new instruments for the observation and capture of bodily movement. Finally, the fifth section, "The Future of Fatigue," briefly considers the divergent outcomes of the physiological investigations of fatigue between the United States, Britain, and France, leading to a confluence of ideas at the close of one era of physiological research and the beginning of another, in 1947.

Physiologists, Sociologists, and Factory Workers: The Problem of Industrial Fatigue

Written in pencil in the summer of 1915, a manuscript with the non-descript title "The Birmingham Story" found its way to Hugh Macdonald Sinclair, a biochemist and expert nutritionist at Magdalen College, Oxford.

The manuscript, signed "C.S.S.," seemed to be a quasi-ethnographic account of a period of factory work. With flashes of vivid prose, it captured a complex, chaotic symphony of metal and movement, yet also familiarity and repetition. This passage in particular stands out:

> A scene of Titanic activity stretches around. Motion and power in the great distributing shafts and pulley wheels spinning so smoothly that but for a flicker in the light reflexed from the oil on them they might seem at rest, and in the strange half human movements and reversals of the machine lathes and their tools. And feeding, directing, tending, using, this frenzy of steel and same intent figures of men, and women, in the same places, in the same attitudes, as a month ago. There is the same man cutting the seemingly same groove band with the same guide turn of the lever as a month back. The same French-woman— does she look a trifle paler and more tired?—twisting and untwisting the pin of a burnish facer just as she did all the time last month.

What would all this movement and effort be for, one might wonder? He continues:

> Seventy hours a week. The same all-pervading reek of oil and the same trolley-boys threading the narrow gangways collecting the finished pieces from the living inanimate and animate machinery[.] 45,000 3 in. shrapnel shells a week. (Sherrington, in Sinclair 1984, 97)

The "C.S.S." is Charles Scott Sherrington who would be knighted in 1922, and win the Nobel Prize for "Physiology or Medicine" ten years after that. He was the author of numerous papers on the neurophysiology of reflexes, as we saw in chapter 1, and in *The Integrative Action of the Nervous System* (1906) he famously described the "reflex arc." This is a shortcut in the nervous system of even primitive vertebrates that allows bodily reflexes, such as withdrawal of a limb from pain, to occur far faster, since in those cases sensory nerves bypassed the brain stem and activated motor neurons in the spinal cord. More immediately pertinent to his factory visit, in that book Sherrington first defined "proprio-ception," where receptors positioned throughout the muscular tissues and accessory organs (tendons, joints, blood vessels) provide feedback to the organism on its posture, position, and movement

(Sherrington 1906, 129–30). After cycling from Oxford to Birmingham in 1915, by then a well-established physiologist and one of the pioneers of neurophysiology, he volunteered to spend three months working in the VickersMaxim munitions factory. At that factory, and many others like it, workers were producing shell casings during a period of heightened production in World War I, often laboring twelve hours a day, seven days a week. The majority were women, working to keep infantry supplied at the front. Working alongside a mixture of men and women from across the country and, indeed, the European continent, he was to conduct research on industrial fatigue on his fellow munitions workers. The insatiable need for high-quality munitions necessitated a set of precise movements in the production line for each worker, and therefore a rigorously scientific observation of their bodies. For each shell casing, a large number of operations were necessary. As Sherrington himself explains with his newfound factory experience: "From ingot to finished state each shell-case passes more than sixty separate operations, and after each process the gauge decides whether so far it is up to standard" (Sherrington, in Sinclair 1984, 95). After a series of highly repetitive movements, a measuring gauge decides the suitability of the shell for either the front line or the scrap pile.

Between 1914 and 1915, the working week for a factory worker was typically seventy-five to eighty-five hours per week, ninety hours was not infrequent, sometimes reaching one hundred. Meanwhile, the British government was aware in the wake of the Boer War (1899–1902) of the high rate of rejection of potential army recruits because of health and physique, courtesy of the *Report of the Interdepartmental Committee on Physical Deterioration* of 1904 (McIvor 1987, 161). By the end of World War I, there were no less than 13 million workers across Britain in munitions factories alone (Gillespie 1987, 245). The working conditions, including the long hours and the poor state of physical health of the workers, led to an outwardly measurable variability in terms of the quantity and quality of hourly output, called the "industrial work curve" (Myers 1925, 907). The Health of Munitions Workers Committee (HMWC) was set up in 1915 to look into this correlation between a newly identified object of physiological inquiry, fatigue, and the already quantifiable outputs that varied according to particular shifts and times of day, as well as individual productivity levels. The organization of industrial work, prior to the rise of psychology in the workplace, resided squarely within the realm of physiology at that time: "What had once been the subjective feeling of the industrial worker was now objectified into a scientific entity and social problem, the solution to which depended on the intervention of physiologists," claims Gillespie (1987, 239). The work of the HMWC continued through its reincarnation as the Industrial Fatigue Research Board (IFRB), which was first convened in 1919 with Sherrington as chairman. The IFRB was a body of scientists, civil servants, psychologists, medical doctors, and industrialists assembled under the aegis of Britain's Medical Research Board (MRB), the precursor to the Medical Research Council. Under Sherrington, however, the remit of the Industrial Fatigue Research Board, as stated in its inaugural report, was to "consider and investigate the relationship of the hours of labour and of other conditions of employment; including methods of work, to the production of fatigue, having regard both to industrial efficiency and to the preservation of health amongst the workers" (IFRB 1920, 20). A second IFRB report was published two years later, in 1922, of enough significance that it was reported by the journal *Science* (1922), and indeed most reports were regularly reviewed in the *British Medical Journal* (e.g., *BMJ* 1921, 1922, 1924). Generally, IFRB reports followed a similar format, starting with brief descriptions of the experimental tests for efficiency and fatigue, followed by more detailed accounts of the results obtained in various industrial settings relating output to the duration of work shifts and rest periods. Later sections included detailed considerations of environmental factors in the workplace, such as temperature, humidity, ventilation, lighting, and so on. Further aspects of the reports' content are offered, for example, in the BMJ's review of the IFRB report from 1922. Although much of the data were provided as a series of tables, the reports were amply furnished with diagrams that would help make the findings "of direct value to employers of labour, and to welfare workers and factory inspectors," according to the BMJ (1922, 322). The diagrams "show at a glance the hourly and daily variations of output observed under various conditions, the effect of regular rest pauses on output, the improvement of output caused by more adequate lighting and by better ventilation, and the value of certain psycho-physiological tests in measuring the skill of compositors" (322). In other words, much of the data were made immediately legible in graphic terms, as a series of optimal or suboptimal industrial work curves. As the work of the IFRB continued, its mission progressively incorporated more attention to the wider environmental factors beyond individual worker fatigue. In 1924 the task of the IFRB was defined in these wider terms by the BMJ:

> To advise and assist the Medical Research Council in promoting scientific investigations into problems of health among workers, including occupational and environmental factors in the causation of ill-health and disease, and the relation of methods and conditions of work to the functions and efficiency of body and mind; and in making known such results of these researches as are capable of useful application to practical needs. (BMJ 1924, 349)

As the mission became more expansive, it should come as no surprise that the IFRB would change its name and widen its remit once again in 1927, to become the Industrial Health Research Board (IHRB) (see Schilling 1944). Despite the reorganizations and name changes, these boards drew upon a common pool of researchers and had common means of dissemination, mainly scientific monographs using highly technical language. Increasingly, however, articles were written for the employers, the welfare workers, and the factory workers, to address an emerging specialization within publishing, that of management and management science.

The IHRB went beyond the issue of fatigue, and started to take wider questions of occupational health and environmental conditions for workers seriously (Gillespie 1987, 257). The work of both the IFRB and then the IHRB involved fieldwork in factory settings as well as laboratory research. This assembly of professionals, an alliance between science and industry, was rather uniquely tasked with the improvement of the health, and not only the productivity, of factory workers, and reached its peak of efficacy through the 1937 Factories Act, which ruled on acceptable working hours, ambient conditions, and rest breaks. Along the way, encounters with various unions, and the need to recruit female members for the board because of criticism from women's groups including the National Union of Women's Suffrage Societies, meant that there was a socially progressive bent to this British research. As discussed further below, this was to be contrasted with contemporaneous industrial research in the United States. Frederick Wilmslow Taylor emphasized the stopwatch in his industrial research in the 1880s and 1890s, and famously coined this new orientation "scientific management" (Taylor

1912). Closer to Sherrington's work on fatigue, however, was Frank and Lillian Gilbreth's motion and micromotion studies from 1907 onward. Their stress on the efficiency of the worker's bodily movements was to reduce unnecessary actions, charted through systematic observation and the use of chronophotographic and other cinematic techniques in the laboratory, starting with bricklayers (Gilbreth 1906), factory workers (Gilbreth 1911, Gilbreth and Gilbreth 1916), and then, in World War I, munitions manufacture with his "Fundamental Problems Involved in the Manufacture of Munitions" (Gilbreth and Gilbreth 1917).

A total of eighty-four monographs were written by the IHRB and its precursors until 1939, and while the research could never boast a wide or even receptive readership within industry, their work informed the 1937 Factories Act, which addressed health and safety issues and included detailed provisions for the regulation of cleanliness, lighting, temperature, and working hours. Notable in this Act, as for the preceding Acts throughout World War I, was their accommodation of the increasingly large numbers of women in the workplace, at a time when men aged between 18 and 40 were conscripted. Therefore, while much of the IFRB and IHRB language is couched in gender-neutral terms, it should be noted that over three-quarters (in fact, 77.6 percent) of munitions workers were women (Pencavel 2014). As Florence argues in his *Industry and the State* of 1957, the IHRB's work on fatigue and conditions chimed with the trade unions' demand for a forty-eight-hour work week. Although this had more or less already been achieved by 1919, says Florence, the evidence accumulated by the IFRB and IHRB helped the rollout of these benefits to all workers, so "justified the 1937 Factory Acts' forty-eight hour normal ceiling for women and young persons in all manufactures" ([1957] 2003, 54). Although the Factory Acts had started in 1833, even earlier in 1819 for legislation relating to cotton mills, the increased scientific credibility of the evidence put forward by the IFRB and IHRB guaranteed not only shorter working hours for more workers, but also the creation of more frequent rest breaks during the day, as proof mounted that this actually aided the industrial work curve.

After 1937, the IHRB "turned to the psychological and social factors underlying efficiency and morale," claims Florence (54). The particular relationship between science, industry, and the state had intensified throughout the First World War, so that by the Second World War factories were required to appoint welfare officers and consultative bodies. To this end, the IHRB published a series of pamphlets in order to disseminate the scientific research into a more pithy, easily readable form for consultants and the new management class. Pamphlet No.2, *Absence from Work: Prevention of Fatigue* (IHRB 1944), for example, explained that fatigue was responsible for the high levels of absenteeism, a highly counterproductive outcome in a time of war. According to the IHRB research, absenteeism levels were roughly 6–8 percent for men and 10–15 percent for women, clearly affecting optimal munitions production. Blame could not be solely attached to the individual worker, and so the role of high-level management was crucial in tempering these figures. The newly established *British Journal of Industrial Medicine* reported on the IHRB findings in their inaugural volume of 1944. Even in a time of maximum war effort, the IHRB recommended: "Weekly hours of work should not exceed 60 for men and 55 for women . . . for heavy manual workers and women with home duties these hours are still too long . . . a week-end break is important . . . weekly hours should whenever possible be spaced over a five-day week . . . in severe muscular work a break of 5 or 10 minutes every hour is better than a single break of 15 minutes in the middle of a spell. . . . It is often possible to adjust the machine, the seating, or arrangements of work so as to avoid excessive strain and waste of energy" (IHRB 1944, in D. S. 1944, 201). Fittingly, the *British Journal of Industrial Medicine,* which started in 1944, and seemed to be a summation of the work from the HMWC, the IFRB, and then the IHRB, itself changed its name to *Occupational and Environmental Medicine,* but only in 1994.

Curves and Lines: Mosso, Ergographs, and the "Curve of Fatigue"

We have already alluded to the industrial work curve, and also to the instruments employed by physiologists in the factories to compile reports on how various nutritional and environmental conditions affect fatigue and energy expenditure. The genealogy of those instruments arguably extends from pioneering experiments in the 1850s on frog muscles. But fatigue as a distinct problem, to be defined and measured, would not occur until decades later, and involved multiple modifications of the equipment in order to be measured.

In his 1891 book *La Fatica,* first translated into English in 1904, Angelo

Mosso saw the potential after Galvani and Faraday in applying electrical stimulation to the body in order to study muscle function and fatigue. Born into a humble background, he had observed first-hand the social injustices meted out to the common laborer. He was "distressed by the detrimental effects of work on the bodies of children, sulphur miners, and Sicilian farmers and by the physical changes experienced by immigrants seeking work in the United States" (Di Giulio et al. 2006, 53), and believed that physiology could help. His physiological exposure to fatigue was instigated by visiting Hugo Kronecker's laboratory in Leipzig in 1873. Kronecker had been conducting experiments on muscles extracted from frogs for several years, based on initial experiments with frogs and recording equipment first devised and assembled by Helmholtz in the 1850s. From observing these isolated muscles in Leipzig, however, Mosso was impressed with how Kronecker "succeeded in obtaining 1,000 and even 1,500 contractions, one after the other, with the greatest regularity" ([1891] 1904, 81). With such regular contractions in any single muscle, naturally one would expect an alteration in the amplitude of the contractions over time as the fatigue takes hold, which, if represented graphically, would be a curve. In his "Über die Ermüdung und Erholung der quergestreiften Muskeln" (On the fatigue and recovery of striated muscles) of 1871, Kronecker, following Helmholtz's methods and findings, had formulated what he described as a "curve of fatigue" as his first law of fatigue: "The curve of fatigue of a muscle which contracts at regular intervals, and with equally strong induction shocks, is represented by a straight line" (in McComas, 2003, 65). The expected "curve" turns out to be a straightforward arithmetic progression for Kronecker, what would be graphically depicted as a straight line, which Mosso found puzzling: "As the contractions follow one another, their height diminishes in proportion as the fatigue increases, and goes on diminishing with regularity until it disappears altogether," as Mosso puts it ([1891] 1904, 81). Mosso would open this up for reinvestigation. Kronecker's second law was more intuitive, and relates to a modulation of the "curve": "The height of the contractions diminishes more rapidly the more rapid is the rhythm in which they are produced, and *vice versa*" (in McComas, 2003, 65). In other words, the level of fatigue is dependent on the frequency of electrical stimulation, and thus on the frequency, and duration, of contractions and extensions in muscular activity. Were there to be an instrument that could record each of the muscular contractions over time, and graphically depict the declines in amplitude, such curves—or lines—of fatigue would be rendered visible for that particular muscle for all to see, whether frog, dog, or human. Furthermore, such a device might definitively answer one of the prevailing questions of the time: Is fatigue purely the result of the buildup of toxins in particular muscles, or does it reside purely in the mind?

The instrument that Mosso developed was the ergograph, from the Greek *ergon*, meaning "work" (see Figure 20). It was designed to follow up on Kronecker's, and by extension Helmholtz's, laboratory research, who both extracted and isolated a single frog muscle outside of its musculoskeletal trappings and cerebral influence. The ergograph required no such separation between muscle and surrounding body, yet also recorded the amount of work (application of force) a muscle achieves in its course of contraction and extension, and by means of an attached stylus was able to chart the progressive rate of fatigue in the subject over time. In a chapter of his book on the measurement of mental fatigue of 1908, Abraham Abelson straightforwardly describes the ergograph and its mechanism of graphical inscription:

> This instrument is composed of a thread at the end of which a weight of two or more kilogrammes is attached. At the other end, the middle finger of the right hand is tied at the second phalange, care being taken to isolate the working muscles of this finger completely by having the arm and all the other fingers fixed. The middle finger is then made to lift the weight as many times as possible until it can no longer move. The subject is made to contract his middle finger at regular intervals of time, generally every two seconds. A point is also arranged which shows the tracings of the movements on a cylinder covered with sooty paper. As the weight is raised, so the point traces a mark on the cylinder. When the muscles of the finger are quite exhausted a level line is shown on the tracing. (Abelson 1908, 352–53)

Before the muscles are exhausted, however, the highest points of the tracing can be joined up, and as we know from Kronecker, "the line thus obtained is of characteristic form and is known as the Fatigue Curve" (353). Repeating Kronecker's work with this new instrument, Mosso found that "we have a great change from the straight line found

FIGURE 20. *Mosso with his desktop ergograph at Turin. Photograph: Archive of Professor Marco Rodolfo Galloni, University of Turin.*

by Kronecker as the expression of fatigue in frog's and dog's muscles after separation from the body. This shows that in man the phenomenon is considerably more complex" ([1891]1904, 91). Indeed, having muscles remain in situ, nested among other muscle and nerve tissue, and with some as-yet uncharted but undeniable connection with a cerebrum, the relationship between mental fatigue and muscular fatigue was now a further issue to investigate experimentally. Even if hundreds of charts showed the curves of fatigue for frogs and dogs, how could muscular fatigue for human subjects be measured in the workplace without extracting specific muscles from the body? Furthermore, what was the influence of mental processes on this fatigue curve, and could this factor be isolated?

One of the chief areas of interest for Mosso and Abelson with the ergograph was therefore the influence of mental fatigue on these physical traces. Previously, basic experiments had already tried to ascertain whether fatigue was a phenomenon local to the muscles, or produced in central brain areas—that is, was a phenomenon of the "center" or the "periphery." As Josephine Goldmark discusses, in 1865 the physiologist Ranke injected fatigued muscle tissue from one frog into a second, unfatigued frog. The second frog showed signs of fatigue. In a paper of 1890, Mosso replicated the experiment on dogs, and concluded that the noxious substances that accumulate in the blood as a result of fatigue are not restricted to local body parts (Goldmark 1912, 15). These experiments did not account for the role of the brain in fatigue, and whether mental fatigue affected the curve of fatigue. The separation of the muscles of frogs and dogs from their bodies told one story of fatigue, but as we saw in Mosso's answer to Kronecker's straight line of fatigue, human subjects with brains as well as muscles were indeed more complex. This is why Mosso added another innovation to the bench ergograph experiments—namely, the electrical stimulation of the finger doing the work so that, in effect, it mimicked the decorporealized muscles of the frog:

> To eliminate the mental element which might alter the fatigue curve of the muscle, I thought of stimulating the nerve of the arm, or rather the flexor muscle of the fingers. . . . Fatigue was produced with the same curve as when the muscles contracted voluntarily. . . . One must conclude that the mental factor does not exercise a preponderating influence, and that fatigue may even be a peripheral phenomenon. (Mosso [1891] 1904, 101)

As Edwards et al. (2013) observe, this question of center and periphery in fatigue was actively debated by physiologists well into the twentieth century, by Bainbridge in his *Physiology of Muscular Exercise* (1931), for instance, and Merton in a paper of 1956, "Problems of Muscular Fatigue," who settled the matter by showing that recovery from fatigue required the restoration of the circulatory oxygen supply.

Every set of muscles on every human subject was different, of course, and so Mosso's ergograph would offer a profile for each individual experimental subject, and chart in easily graspable ways their capacity for repetitive work. "The ergograph thus gives us a record of one of the most intimate and most characteristic features of our individuality— the manner in which we fatigue," thought Mosso ([1891] 1904, 92), and by and large this factor remained constant. "The curve is constant for the same person provided he is in the same condition of health and repose," explains Abelson (1908, 353), but Mosso found that variations in levels of mental fatigue, or the ingestion of particular foods or stimulants, was ripe for exploration via this method. Although the original and ideal application of the ergograph was for the measure of fatigue for an individual subject, as we will see with Amar below, the flip side was that the same equipment could measure the amount of work undertaken by each individual. An experimenter like Mosso, who built upon Marey's original graphical method, desired only to compile rigorous fatigue profiles of individual workers based on the measurement of muscle fatigue in the handling of a given tool. It would be a short step, however, to use that same experimental apparatus against the worker, as it were, to measure and grade the effectiveness and efficiencies of the worker in terms of actual work (effort, movement) undertaken. This is a fundamental difference in orientation between Mosso and Amar, as well as Gilbreth in the United States. Incidentally, the graph of progressive muscular fatigue shown by Mosso's ergograph in 1891 is mirrored in almost uncannily similar ways through an entirely different technology at the University of Oxford almost a hundred years later in a 1978 paper in *Nature* through the use of nuclear magnetic resonance (NMR) spectroscopy, which also measured fatigue in muscles from frogs and then human forearms.

The ergograph did not arise fully formed out of nowhere, of course. Its precursors included a veritable constellation of instrument designs and their variants in general circulation among European physiologists in the nineteenth century, evidently a boom time for instrument makers such as Charles Verdin and Louis Francois Clément Bregeut in France, and Charles Wheatstone in England (see, e.g., Schmidgen 2014). In 1850 Helmholtz had been the first to precisely measure the movement arc of the isolated frog muscle using a "frog-drawing-machine" *(Froschzeichensmachine)* he had developed between 1848 and 1852, and that later became known as the myograph (Norton Wise 2010). From this and other data, Helmholtz could extrapolate the speed of signal propagation in nerves, the first scientist to do so. The myograph was the instrument that Mosso directly adapted in 1890 into his ergograph (Abelson 1908, 352). Helmholtz's myograph attached an isolated frog muscle to a hook and, once stimulated through electrical induction, the contracting muscle lifts a metal frame with a stylus attached. The arc of this cycle of muscular contraction and extension is then drawn as a curve onto a rotating drum, what Helmholtz described as "the curve of the *Energie* of the frog muscle" (Norton Wise 2010, 97). He was able to record the conduction velocities of motor nerve fibers in the gastrocnemius muscle of a frog's leg at around twenty-seven meters per second. As we shall see, Helmholtz's interest in generalized laws of energy transfer, stimulated by Carnot's diagrams of steam engines and heat exchange, certainly frames the muscular movement in terms of work. An illustration of a cross-section of the *Froschzeichensmachine,* reproduced as a plate in a later publication of 1883, similarly casts the whole apparatus in this light, thinks Norton Wise: "Helmholtz's device treats the contracting and relaxing frog muscle quite literally like an engine burning fuel to produce work" (96). Indeed, any point on the curve of contraction and extension inscribed on the drum is also a calculation of the net work achieved by the muscle, as Ludwig observed in 1852 (97). Helmholtz's myograph effectively displayed the work curve of an individual muscle. Marey's modification of the myograph, as illustrated in *La machine animale: Locomotion terrestre et aérienne* (1873), eliminated distortion and allowed the recording of successive movements as a series of curves on the drum, "permitting a comparison of the intensity of contraction in a series of stimulations," as Mosso explains ([1891] 1904, 78). This innovation effectively turned the work curve of Helmholtz into the curve of fatigue of Kronecker and Mosso.

Meanwhile, the sphygmograph measured blood pressure, and was originally invented by Karl von Vierordt in 1854 as a rather cumbersome wood-framed device in which weights were added to offset the pulse and therefore produce an imprecise estimate of pulse strength. Marey improved upon it in 1863 not only by making it portable, and so wearable on the forearm, but also by attaching a springed arm that magnified the pulse and graphically inscribed it onto paper (see Figure 21). The configuration of this apparatus in terms of a mechanism in springed suspension that directly responded to the naturally produced forces on the skin surface of a rhythmic pulse, and then output the pulse as a graphic inscription, was a significant step toward Mosso's ergograph. Mosso's own sphygmomanometer made by Verdin in 1895, and described in his paper "Sphygmomanomètre pour mesurer la pression du sang chez l'homme" (Mosso 1895), could measure blood pressure this time through the hands, and like Marey's version of the myograph could output onto a rotating writing drum to display pulse over time.

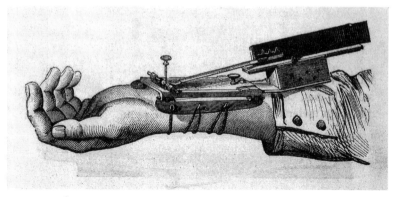

FIGURE 21. *Étienne-Jules Marey's portable sphygmograph. Public domain image, courtesy of the Wellcome Trust (L0012232).*

Other conceptual vectors and engineering configurations were involved in the esthesiometer, which in its most basic form measures skin sensitivity through the discrimination of two points placed on the skin surface of a subject. While tactile discrimination and the perceived two-point threshold was seen as a map of tactile sensitivity within different regions of the body, discriminative ability was used as an index between human subjects by Weber and then Fechner (as described in detail in chapter 2). Binet and Henri's (1898) unforeseen experimental results demonstrated that such discriminatory ability, and therefore tactile sensitivity, varied if the experimental subject was experiencing fatigue. The instrument maker Charles Verdin's version in 1890 involved a long horizontal beam with two vertical sliding rods, each of which ended with a rounded ivory point. Several iterations of this instrument were constructed by Verdin and made available, including a handheld sliding esthesiometer and a more adjustable model with springs he called a "dynamometric esthesiometer" (Verdin 1890, 65). For all variants, the distance between the two vertical points was easily measurable because a millimeter scale was printed on the horizontal beam. Von Frey used a variant in his experiments of 1894–96, terminating the rods with fine hairs he called "Von Frey Hairs" (see Figure 22). Jastrow employed a simplified version without the spring in his experiments from around 1893 onward, and then a more refined version in 1910. Münsterberg's esthesiometer was impressively sophisticated, on the

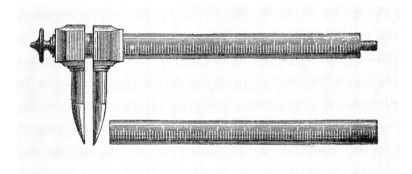

FIGURE 22. *The esthesiometer as depicted in Lombroso's* Criminal Man. *Public domain image, courtesy of the Wellcome Trust (L0022895).*

other hand, consisting of a flexible rod with a handle at one end and a contact point at the other, to which various "contact-pieces" could be inserted. These might include "points arranged in various figures, continuous squares and circles, and surfaces of various sizes, for investigating the thresholds of sensation and difference," explains Baldwin (1905, 612). In fact, temperature sensing could also be tested by attaching a set of small pans to the contact pieces. Cesare Lombroso, the pioneer of criminology, used his variant of the esthesiometer for the classification of criminals, mapping the lack of sensitivity of certain subjects (prostitutes and peasant women, for example) in *L'Uomo delinquente* (Criminal man) (1878). Esthesiometers, which involved setting points on a slide in order to measure thresholds of tactile and pain sensation, and algometers, an instrument that measured sensitivity to pressure or pain on the skin, were both designed to "standardise and quantify the intensity of a pain-producing stimulus" and "to map sensitivity on body surfaces, to study biosocial difference and to measure analgesic effect," explains Noémi Tousignant (2014, 112–13). Meanwhile, working on a series of papers with collaborators, Alfred Binet had been interested in esthesiometry and its application to mental fatigue for several years (e.g., Binet and Vaschide 1887; Binet and Courtier 1897). With Victor Henri in 1898 he coauthored a synthesis, the first monograph of its kind, on mental fatigue (Binet and Henri 1898). Their purpose was to give teaching and education a solid experimental background and to clarify what still remained the rather imprecise notion of "mental fatigue." To do this, they used a variant of the esthesiometer in order to assess the relationship between mental fatigue and tactile discrimination. Binet and Vaschide used Mosso's sphygmomanometer to see whether intellectual work had any effect on blood pressure, which it did. Then Binet with Henri employed Mosso's ergograph, modified according to Binet's advice, to assess subjects doing sustained intellectual work. Unsurprisingly, they found: "In summary, intellectual work seems to modify muscular strength; moreover, this modification differs depending on whether the work is short or sustained, and whether it is linked with a particular emotional state or not" (Binet and Henri 1898, as translated in Nicolas and Makowski 2016, 19).

Movement and Fatigue: Amar and the Worker as "Human Machine"

The truly scientific investigation of the workplace, in particular the industrial workplace, remained largely in its infancy when Jules Amar wrote *Le Moteur humain et les bases scientifiques du travail professional* (The human motor and the scientific basis of professional work) in 1914, a year before Sherrington's munitions factory visit. As a result of the intrusion of World War I, its first translation into English was delayed until 1920, and involved a significant but subtle alteration of emphasis in the subtitle, presumably the result of interim developments in the United States: *The Human Motor: Or, the Scientific Foundations of Labour and Industry*. As summarized by a contemporary review in the *New York Times,* "The work is a scientific study and assembling of all the physical and physiological data connected with the efficiency of the human body in industrial work" (1921, 46). Amar was a pupil of Auguste Chauveau, the head of the Marey Institute after Étienne-Jules Marey died. Although there would be ideas and instruments in common with his fellow Frenchman, Marey, and also the so-called "scientific management" of the Americans Taylor and Gilbreth, Amar remained focused on physiological factors of energy and fatigue in the workplace.

"The result of fatigue is a lessened aptitude for work," complains Amar (1920, 206). His means of empirical investigation involved the use of ergographs to trace the effects of fatigue on the body, just like Mosso. Also like Mosso, Amar used the resultant ergograms or "ergographic tracing" as a curve of fatigue, showing the "decrease in the quantity of work done by contraction," clearly shown as a downward curve. Following Kronecker's second law of fatigue, although strangely not explicitly citing it, Amar too finds "the curve will descend quicker or slower according to the rhythm of the contractions and the intensity of the effort" (206). Amar found that the tracings revealed an unusual feature of fatigue, one which Mosso barely considered: its intermingling with pain in our habitual experience. In the laboratory, and especially in the case of isolated frog muscles, "real muscular fatigue is slow," he says, whereas "it is shown very quickly on the ergograph in consequence of the sensation of pain" (206). Amar does in fact refer to contemporaneous physiological explanations of fatigue as a localized buildup of noxious chemicals in the periphery of the muscle, and even to the potential presence of lactic acid. But what shows up in the ergograms is inevitably the result of uncomfortable and even painful sensations that arise within the human subject, and this is what limits repetition of the same range of motion: "The conditions of the ergographic work of the finger generally cause discomfort, and the traces . . . show this discomfort rather than the real fatigue of the flexor muscles" (206). To better accommodate these limitations, to smooth out the work curve, overcoming this discomfort and pain would be the key to efficiency in the workplace, and would have real-world implications.

If the ergograph was based on Mosso's principle from *La Fatica,* he had to modify existing instruments to derive the right measure. In turn, others would adapt Mosso's ergograph for their own purposes. Trèves's "brachial ergograph" from 1898, for example, was based on Mosso's idea (298), where the subject was tested while standing, and movement of the subject's arm turned a crank. Results from Mosso's or any ergograph bench equipment led to the diagram of work rate or overall work efficiency, known as the ergogram. This was inscribed onto paper through a stylus attached to the machine, in a way familiar to those other instruments of measure such as the sphygmograph. While Mosso was mostly interested in the identification and scientific measurement of fatigue in as rigorous a way as possible, some twenty-five years later the Tunisian-born Frenchman Jules Amar wished to broaden the mission of this measurement. For Amar, the ergograph could not only investigate increasing yields in the workplace but also categorize, grade, or otherwise measure the suitability of workers for their tasks. Ergograms could therefore be co-opted into the technics of efficiency. As he states at one point: "The secret of attaining the maximum yield from a workman is to proportion the elements of the work so that the muscles contract without pain or excess of local fatigue" (208). Later he places the value of the ergograph exactly in terms of minimizing the loss of yield: "The principal object of ergographs is to investigate the decrease of work under the influence of fatigue" (299).

Hence a more scientific measurement of labor output is possible, given a time *t,* dividing the work done for a particular duration, to chart the direct drop of output as a result of muscular fatigue. Amar is clearly more allied with Taylor and Gilbreth's scientific management than with Mosso when he makes the assertion: "A scientific organisation of labour must assure that the power of the workman is practically constant at each period, and that it is not reduced by too rapid fatigue, due to bad mechanical or physiological conditions" (307). Ergograms could therefore help determine the influence of other factors on the maintenance of the constancy of the worker's output. Many types of factors, including bodily states altered by the effects of certain foodstuffs or even narcotics, as well as environmental conditions, could then be plugged into the experiment to measure their potential impact on the acceleration or deceleration of fatigue on the total yield:

> Ergograms reveal also, by their shapes and the quantities of work they represent, the influence of the numerous factors of human activity: the nature of this or that food, external temperature, position of the body, the magnitude of the effort to be produced, speed of walking, and the rhythm of the movements. By their means . . . the speed of economical work and, to some extent, of muscular indefatigability, can be determined. (208)

The scientific measurement of labor, via ergographs and other instruments like the sphygmograph, and the presentation of graphic data throughout the book, were all part of an expansive scientific vision wherein the smooth constancy of output should be maximized while the friction of mechanical and physiological limits minimized. For any putative *moteur humain,* fatigue is inherently an internal limit of the physiological machine or the laboring organism that, if we remember, led him to say: "Fatigue can be defined as the effect which limits the duration of work" (206). It prompts Amar to analogize human worker and nonhuman machine in no uncertain terms toward the end of *Le*

Moteur humain, going so far as to state that "Man is indeed a valuable machine" who can be placed in any situation and utilized "so long as his fuel (food) can be supplied" (463). It is in book 2, "The Human Machine," that Amar starts to discuss Helmholtz's ideas, and his position on energy and fatigue becomes an expressly Helmholtzian analysis of the organs and physiology of the body in predominantly mechanical terms. At several stages Amar's language is unequivocal: "The organs of locomotion are the bones and muscles; the former being levers, and the latter being powers" (86). At another point, as if he is describing isolated frog muscles in the laboratory, he likens the contraction and extension of the muscles in human movement to the mechanism of a steam engine: "The contraction of the muscle, and its tetanic [i.e., sustained contracting] character when voluntary, alternates, like the strokes of the piston of a steam engine" (128). Based on these and earlier observations about efficiency and energy, the problem of pain, discomfort, and fatigue are of significance only insofar as they alter the industrial work curve and therefore diminish net output. It should come as no surprise that he ends up equating the worker to a valuable machine, one that must be maintained to keep working as efficiently as possible:

> The maintenance of the human machine is as difficult and important a matter as the maintenance of any inorganic motor. We must do our best to eliminate all internal and external conditions tending to cause depreciation. Especially must physiological discomfort be avoided, and removed, as far as possible. In the due proportion of effort and speed, in industrial labour, lies the key to intensive production and the well-being of the workman. (463)

For the remainder of this section, it is worth retracing points along Amar's path that led him to the seemingly antithetical position that the incremental progress toward efficiency of the "human machine" was compatible with the well-being of the worker.

Both Mosso and Amar had referenced Marey and his graphic method extensively, noting his ingenuity in modifying and improving the instruments of measurement. Both Mosso and Amar used such instruments themselves, especially the ergograph, and referred to other experiments by physiologists who had used, and then modified, this instrument. But another factor they had in common, like many scientists in that period,

was their appreciation of Helmholtz's ideas on the transfer and conservation of energy. Helmholtz's paper "On the Interaction of Natural
Forces" was first delivered in 1854 and published in 1873, and was
hugely influential. In it he coined the term *Arbeitskraft* (labor power)
for forces that operate across organic beings such as animals and humans, with their chemical processes, and machines, with their physical forces. Forces are considered broadly equivalent in their mechanical effect, then, whether they be from animals, humans, or machines,
and Helmholtz identifies a transferability of energy between them. An
ox with its plough, a windmill grinding wheat, or a worker in a factory
were situated in a continual network of energy transfer, where food
and nutrition helped power the organism in the same way as steam
powered the steam hammer, or the wind powered the windmill. The
chemical energy that arises through digestive processes by the ox, say,
is translated into kinetic energy through movement, such as turning a
wheel, and thence into mechanical energy, say the grinding of wheat or
drawing of water. Helmholtz himself refers to the older term for kinetic
energy, *vis viva,* or "living force" (Helmholtz 1873, 157), which permits
ambiguity as to whether its origin is organic or machinic. Robert Mayer
in 1845 had shown that heat arises within the animal as a result of their
movement and is dissipated, although at the time Helmholtz was writing the exact mechanisms were unknown. Amar's debt to Helmholtz's
"On the Interaction of Natural Forces" is therefore the acknowledgment
that the chemical energy in animals and humans alike is converted into
mechanical force and results in heat, and that energy is conserved in
the system as a whole. The debt is compounded in the way that char-
acterizes both the animal and human bodies within the system in
machinic terms. In analogizing forms of combustion in the steam engine with the processes of digestion in the animal body, for example,
Helmholtz argues: "The animal body therefore does not differ from the
steam-engine as regards the manner in which it obtains heat and force,
but does differ from it in the manner in which the force gained is to be
made use of" (183). Amar summarizes Helmholtz's position in terms of
an equivalence of energy across organic and mechanical systems thus:

> A body, therefore, possesses the power to furnish a determined
> quantity of energy.... Energy is thus manifested as an invisible and

universal entity capable of changing its aspect without change of magnitude. It is an invariant, as the geometricians say. (1920, 45)

This is not simply physics, then, nor just physiology, but something in-between. Although physiological interpretations of Helmholtz's *Arbeitskraft* commonly saw fatigue as merely a depletion of energy, it remained so influential that it was revisited almost one hundred years later by Georges Canguilhem in 1947 in his extraordinary lecture "Machine and Organism" (1992), discussed below. Mosso had dutifully discussed Helmholtz's ideas on energy in chapter three of *La Fatica*. But he had identified distinctions between animal fatigue and human fatigue, and his ergograph was designed explicitly around this separation. At one point, Mosso acknowledges that almost all experiments on fatigue so far had involved frog muscles removed from their bodies, and that this was a poor model for human bodies situated in their workplaces: "But with frogs it is impossible to reproduce the normal function of muscles and to imitate the action of a man who is doing mechanical work," he says ([1891] 1904, 83). Amar's reading of Helmholtz's conservation of energy took the machinic language far more seriously, and seemed to justify conceptualizing fatigue in the same manner, as an inconvenient by-product of an inherently inefficient, and regrettably unmachine-like, body.

What, then, of the "motor" in this sustained exploration of energy, forces, and power? In the section "Power of Motors," he discusses the usual metrical equivalences of energy (watts, joules, kilowatts, and kilogram-meters), but shifts easily into the language of "live motors" at one point, meaning animal or human energy. Like a true Helmholtzian, he need not differentiate between the sources of this power, whether animal or human, mechanical engine or electric motor. Electrical motors produce a smoothly continuous quantity of "work or mechanical energy developed in unit time," unlike "live motors" or, as he sometimes terms them, "animated motors," which only offer discontinuous work through standard time. Hence fatigue is unavoidably built into the limits of the animated motor itself—that is, the muscular and neurophysiological limitations of the animal or human body that became extrapolated as the curve of fatigue, as Mosso had found. For Amar, "live motors" or "animated motors" are indeed to be compared, albeit unfavorably, with electrical motors and mechanical engines in terms of efficiency and an overall assumption of equivalency: "The expenditure or input of a motor is its consumption either of fuel in inanimate motors, or of nourishment in animated motors," he says (1920, 79). Moreover, whether animate or inanimate, the rhetorical value of the concept of the motor is its abstraction from the immediate particularities of its inputs (food, fuel) and outputs (force, work). What is motoric is the smooth continuity of the conversion from the one to the other. Like an engine hovering around its optimal speed, or a worker in the middle of a shift, the components are working in concert to foster continuity. "In normal continuous work, without excessive fatigue, the muscular fibre becomes more irritable and the fibres and nervous cells more excitable," says Amar. "This is shown by an increase of force and by a smaller consumption of oxygen" (195). Inevitably, since the work curve is shaped by the curve of fatigue, and shifts are broken up by mandated rest periods or unofficial interruptions, the smooth continuity of output cannot be maintained for long. Amar ruefully acknowledges that this discontinuity of labor due to physiological constraints inevitably affects the overall yield of the work:

> In live motors, the work is not continuous on account of the phenomena of fatigue. . . . It is necessarily intermittent since a day's work is interrupted by a certain number of halts or intervals of repose. Taking the total of the work done, and the total duration of the periods of activity, the quotient of these two quantities gives the useful power of the man or animal. (1920, 79)

The useful power of the man or animal, the yield in the workplace, had been as dispassionately examined before by physicists such as Charles-Augustin de Coulomb, who worked out total work as expenditure per meter-kilogram. This was always a theoretical maximum yield, with no room in the formula for inconstants, including fatigue. Amar's heading in book 2, chapter 2, "Yield of the Human Machine," would not seem to countenance the inevitability of fatigue for the worker. But he does concede that the worker may arrive already in a state of fatigue, having "a high static expenditure" (186)—that is, fatigue as part of their background bodily state. Due to the long working hours and poor nutrition, this was exactly what Sherrington had observed in his Birmingham story. The wider implication of the unsuitability of the workforce in a time of maximum industrial output in war, where inconsistency of output had quality and therefore safety repercussions, could at least be mitigated to varying extents with the Factory Acts in Great Britain.

The unpredictability of such static expenditure, the background state of fatigue, meant that there was an indeterminacy at the heart of the concept, despite the desire to smooth out calculations and curves. Fatigue is periodic, disruptive, and by its very nature a threat to continuity: "Fatigue is not a precise thing, and is difficult to define," admits Amar (182), nested among the myriad graphs and tables. Fatigue is not only the friction within the motor, in machinic terms, but always and inevitably a *sensation* too. Even as sensation, fatigue remained difficult to bracket as a subjective phenomenon, or to measure in quite the way that touch and pain had been grappled with by the psychophysicists. Apart from the densely drawn lines that graphed the net expenditure of work, how else was the sensation of fatigue to be measured, exactly? If the most subjective sensitivities and variabilities involved in tactility could nevertheless be charted by virtue of Weber's "just-noticeable difference" (JND) on the skin surface, the phenomenon of fatigue was less precisely located in the body, and its temporality less immediate. "At the limit at which the human forces are incapable of further exertion, fatigue is more pathological than a physiological state, and the degrees of fatigue, prior to that state, are not susceptible of precise measurement, sensation not having physical dimensions" (182), observes Amar. There is a threshold, a point at which fatigue impinges upon the body of the worker, but this point inescapably marks a decline in muscular performance and net yield. At that point, Amar declares, "the appearance of physiological troubles will mark the limits of fatigue" (183). The worst aspects of the unpredictable periodicity and the inconstant threshold of the sensations of fatigue in repetitive labor could nonetheless be partially mitigated by the "entraining" of muscular movement. With more training or exposure over time, greater efficiency and therefore more work could be achieved: "Normal work 'trains' *[entraine]* the muscles, provoking a greater irritability of their fibres . . . and reducing, perhaps, the period of getting into action which shows itself in beginners by the stiffness of their movements" (204). In entraining the body to the conditions of repetitive labor, the worker's body is simultaneously subject to conflicting forces, both from within and without, both mechanical (forces from outside the body acting upon the tendency to muscular inertia or static expenditure) and physiological (the irritability or excitability of the muscles due to external demands). In essence, for Amar fatigue is never a phenomenon in itself. It may slowly converge above a certain threshold to be felt as a sensation, but by then its pathological effects can be graphically shown as an ergograph or through changes in net work. For Amar, its significance lies only ever in its relation to efficiency, yields, or net work:

> Fatigue can be defined as the effect which limits the duration of work. In the case of inanimate motors, owing to molecular alterations of a physical order, fatigue attacks all parts subjected to stresses, though very slowly. In the case of living motors, man and animals, the fatigue either decreases the intensity of the muscular effort, or reduces the contraction of the muscle. . . . *The result of fatigue is lessened aptitude for work.* (206, emphasis added)

While improvements in techniques, various "tricks of the trade," or what Taylor termed "rule-of-thumb" methods to reduce fatigue and increase efficiency on the production line can be passed down by one generation of workers to another, Frederick Wilmslow Taylor, Frank and Lillian Gilbreth, and Jules Amar were all interested in putting the study of the workplace on a truly scientific footing, although very differently from Weber or Sherrington. What Taylor offered, for example, was put in disarmingly benign terms as "the substitution by our workmen of scientific for rule-of-thumb methods" (1912, 16), and their scientific findings would be equally applicable to workers of any gender, age, or level of physical strength. The objective of Taylor's "scientific management" was "to secure the maximum prosperity for the employer, coupled with the maximum prosperity for each employé," he wrote (9), and in fact "maximum prosperity can exist only as the result of maximum productivity" (12). Amar's views diverged from Taylor and Gilbreth, however, in that his concern with efficiency was not tied into the prosperity of the factory owner or business, but more aligned with the health and wealth of the worker: "Man is indeed an automatic engine, but science can improve its working and can exhibit its beneficent effects in increased wages and improved bodily health" (1920, 463). The kinds of scientific principles used in scientific management to increase productivity, it turns out, could also be applied to minimizing the waste of human energy in production. This, for Amar, was almost a matter of moral hygiene: "In truth, it is an incredible thing that there should exist a science able to restrain or suppress the waste of human energy, and that its sovereignty is not yet extended to all its domains, that its beneficial power is not employed to the profit of society and to the advantage of the workers" (1920, 465). Indeed, despite his emphasis on efficiency and the disquieting analog of the human motor, Amar's scope for the science of labor was potentially transformative for the worker themselves. It would be transformative for the worker in a different way when the Gilbreths and Taylor set to work with a whole set of other instruments to observe, measure, and capture movement through graphic methods.

The Age of Measurement: Motion Study and Scientific Management
"This is the age of measurement. The motion model is a new device of measurement" (Gilbreth and Gilbreth 1917, 97). So commences a paper by Frank and Lillian Gilbreth, first presented at the American Association for the Advancement of Science. Three years before Amar's *Motor humain,* Frederick Winslow Taylor's *The Principles of Scientific Management* and Frank Gilbreth's *Motion Study* were both published in the same year, 1911. The attention to time and the speed of the operations of a worker in Taylor's scientific management, what Taylor referred to as "time studies," was complemented by the "motion studies" developed by the husband-and-wife team of Frank B. and Lillian M. Gilbreth, a variety of procedures for the description, systematic analysis, and means of improving work methods. Gilbreth's *Motion Study* involved plentiful examples from observing and photographing bricklaying because his original research involved observations on reducing fatigue and improving processes in his own bricklaying contracting business, published as *Bricklaying System* (Gilbreth 1906). It took a while before the implications of *Motion Study* were perceived as applicable to industries outside of building construction. In that book, Gilbreth considers the habits, health, nutrition, size, strength, skills, temperament, and training of the worker as variables that can lead to efficiency. Gilbreth identified three stages in motion study: first, discovering and classifying the best practice; second, deducing the laws that make it the best practice; and third, using these laws to standardize practice to increase

efficiency. Meanwhile, Taylor's need to establish management as a "true science" immediately stressed its applicability to a number of areas, and he intended his book "to show that the fundamental principles of scientific management are applicable to all kinds of human activities, from our simplest individual acts to the work of our great corporations, which call for the most elaborate cooperation" (Taylor 2012, 7). Although his examples were drawn from the worlds of industry and manufacturing, those same scientific principles could be applied elsewhere, including "the management of our homes; the management of our farms; the management of the business of our tradesmen, large and small; of our churches, our philanthropic institutions, our universities, and our governmental departments," promised Taylor (8). Indeed, Lillian Gilbreth would write a book on reducing wasted effort and therefore fatigue in domestic labor in her book *The Home-Maker and Her Job* in 1927 (L. Gilbreth 1927). Her husband Frank's classic *Motion Study* (1911) was based on his previous article series in the journal *Industrial Engineering,* and the journal's editor wrote the introduction, where he explained that the complete series had engaged the interest of trade and industry from around the world, including "from the iron and steel industry, from the shoe manufacturing industry, from book-printing and bookbinding establishments, and many other industries" (Kent, in Gilbreth 1911, xiii). However, Gilbreth's focus at that point was on efficiency rather than fatigue, and so fatigue was considered a by-product of the inefficiency of motion. This inefficiency could nonetheless be reduced through his principles of "scientific management." It was in 1916 that both Gilbreths would collaborate and publish *Fatigue Study: The Elimination of Humanity's Greatest Unnecessary Waste.* There, they claimed that motion study did not have the dehumanizing effect of scientific management approaches such as Taylor's, and instead in the drive to reduce the deleterious effects of fatigue, fatigue study facilitated the cooperation between worker and management through optimizing rest periods at the same time as retaining one of the fundamental tenets. Many of the principles from *Fatigue Study* remain in use in the twenty-first century in ergonomic design, and are stipulated in the U.S. Occupational Safety and Health Administration laws. Clearly there were methodological approaches in common between their fatigue study and the studies of time and motion, but their usual emphasis on waste and inefficiency of movement was now directed toward minimizing fatigue effects on the worker:

> Fatigue measurement, as applied to the industries, is a new science. It is being developed through a study of the data of activity. The methods of measurement of activity are motion study, micromotion study, the cyclegraph, the chronocyclegraph, and the penetrating screen. Through the data derived by these, we standardize motion paths, motion habits, and all other motion variables. These enable us to test and classify, select and place, both work and workers, and to eliminate unnecessary fatigue. (Gilbreth and Gilbreth 1916, 131)

The "penetrating screen" mentioned was a cross-sectional screen, a superimposed grid upon a photograph that helped determine motion against a multiple exposure image (Lalvani 1996, 159) (see Figure 23). This allowed the Gilbreths to measure the distance of movements and to gauge velocity. It was used "for recording the three dimensional paths and speeds of even the smallest unit of activity" (1916, 122). The cyclegraph, and later the chronocyclograph, were more technically complex apparatus that similarly allowed a precise analysis of work movements. Gilbreth and Gilbreth in their subsequent collection of papers *Applied Motion Study* (1917) demonstrated "that the subdivision of the motion cycle is the important element," and so the "motion cycle can be accurately recorded, hence analysed into elements that may be standardized and synthesized into a recorded method" (1917, 117). In a manner that would be familiar to students of Marey or Muybridge, the result is a trace of the movement habits of a worker, an instantly graspable visual depiction of motion economy. Rather than photographically capturing movement qua movement, however, the resulting cyclegraph depicted the units of a movement cycle—that is, showed how one of a number of "concrete" repetitive movements required of the worker in the context of their workplace was accomplished, and the most efficient movement arc for that unit. It was therefore effectively a diagram of muscular effort:

> The device for recording the path of the motion consisted of a small electric light attached to the forefinger or other moving part of the

FIGURE 23. *The "penetrating screen" of the Gilbreths to measure the distance of movements and to gauge velocity, useful for recording the three dimensional paths and speeds of even the smallest unit of activity. Photograph from Gilbreth and Gilbreth (1917, 86), courtesy of SpringerNature.*

body of the worker. The worker performed the operation to be studied, and the path traversed by his hand was marked by a line of light. An ordinary photographic plate or film was exposed during the time that he performed the work and recorded the motion path described by the light as a white line, something like a white wire. A stereoscopic camera enabled one to see this line in three dimensions. (83)

As opposed to the cyclegraph, the chronocyclegraph cut up the movement through multiple successive exposures, just like Marey's chronophotography, adding the crucial element of time. This opened up to the observer the relative efficiencies of certain motion paths over others in terms of expended effort over time. The single trace of the white line of the cyclegraph became a linear sequence of white dots and dashes in the chronocyclegraph. If the virtue of the cyclegraph was to provide a "simple, easily understood record of the path that any activity followed," the chronocyclegraph was more suitable "when the activity is complicated, and when the time and direction of the elements of the motion must be visualized continuously in order to analyze, measure, synthesize, and standardize the process," they claimed (1916, 122). Furthermore, there was the possibility of adding further lights to different parts of the worker's body, such that multiple synchronous motion paths could be traced, and more of the motion and effort of the worker's body could be captured:

> The chronocyclegraph method enables us not only to see the path of the motion, but also [to see] its directions, and the duration of the entire motion and of its elements. These chronocyclegraphs are made by attaching lights to the moving parts of the body, or machine, as in the cyclegraph, and by introducing a properly timed, pulsating interrupter in the circuit, which may be adjusted not only to record the time and duration, but also to record these with different graphs, representing the paths of each of several motions made by various parts of the body and their exact distances, exact times, relative times, exact speeds, relative speeds, and directions. (121)

The Future of Fatigue in an Age of Machines

A few years later, the brief but intense flowering of the Harvard Fatigue Laboratory from 1927 to 1947 neatly parallels the extramural physiological research that began this chapter. It was initiated by Elton Mayo, a professor of industrial research who ended up at Harvard Business School, and Lawrence J. Henderson, a theoretical biochemist and physiologist who became professor at the Medical School in 1919. As a scientist, Henderson was influenced by fellow physiologist Claude Bernard's concept of the *milieu intérieur* (see chapter 2). He also had wider interests, being so impressed by the work of Italian sociologist Vilfredo Pareto that the Harvard University Press published Henderson's *Pareto's General Sociology* in 1935. Meanwhile, during World War I, Mayo had studied shell shock with a team of physicians, and realized he wanted to study "the adaptation of the normal individual to industry." Mayo went to Paris to study with Pierre Janet in order "to state more precisely the personal and social maladies of our modern industrial civilization," according to a tribute to Mayo in the *Harvard University Gazette* of 1949

(Horvath and Horvath 1973, 19). Mayo and Henderson teamed up to re-search industrial hazards in the 1920s, a period when Taylor's scientific management and the Gilbreths' time and motion studies meant that industry could no longer ignore human factors and ergonomics. From around 1925, Mayo in the Harvard Business School, and Henderson then affiliated with the Massachusetts General Hospital, started to organize research on the psychological problems of industrial personnel. This was funded by the Rockefeller Foundation, who in their annual report of 1930 acknowledged that since 1925 Mayo and Henderson had been directing research with financial support from the foundation on "the psychological factors which control human behavior [in order to form a basis for understanding problems] in business administration, and particularly in the labor field" (Rockefeller Foundation 1930, 167). This funded work led directly to the founding of the Harvard Fatigue Labo-ratory in 1927. The purpose of the laboratory for Mayo and Henderson reflected both the scientific and the wider sociological premise that "the group psychology, the social problems, and the physiology of fatigue of normal man must be studied, not only as individual factors in determin-ing physical and mental health, but more especially to determine their interrelatedness and the effect upon work," in the words of Horvath and Horvath (1973, 20). The extramural nature of this psychological and physiological research involved embracing the wider environmental factors involved in the workers' adaptation to stress. Researchers from the laboratory were assigned to the construction of the Hoover Dam in 1932, for example, and an Ohio steel mill in 1934. Work inspired by Mayo and Henderson would be conducted later in 1939 in situ at the Hawthorne Works of the Western Electric Company, which used fac-tory floor employees, as well as their supervisors, as research subjects. This study was published as *Fatigue of Workers* by the National Research Council (NRC) Committee on Work in Industry in 1941 (NRC Commit-tee on Work in Industry 1941). Although various Rockefeller Foundation reports characterized the research they were funding as "industrial hazards," and Mayo's role was as director of psychological investigation "in certain large industrial establishments" (1930, 222), the kinds of en-vironmental conditions that led to accidents, including disease, were still of interest, and a professor of social anthropology named Lloyd Warner joined the team specifically "to determine the social factors contributing to and resulting from industrial hazards" (Horvath and

Horvath 1973, 22). In parallel, others in the laboratory were conducting groundbreaking research on the then-burgeoning field of exercise physiology. This involved the first use of a treadmill with calculated gradients from 1928 for measuring oxygen uptake in athletic subjects, a comparison between treadmills and bicycle ergometers, and a series of papers from 1931 entitled "Studies in Muscular Activity" (Horvath and Horvath 1973, 105–6). During World War II, the laboratory pursued this exercise physiology work in field studies, assessing nutrition, fatigue, and changes in lung capacities within extreme environments. A team led by David Bruce Dill observed Peruvian natives at high altitudes in the Andes for six years, for example, and the NRC's *Fatigue of Workers* report highlighted the significance of these data during wartime for deployments in tropical environments, and indicated some problems in aviation medicine. In its review of the NRC report, the *Journal of the American Medical Association* acknowledged the report's value in revealing the complexities of industrial fatigue, through its corralling of expert scientific testimony: "It gives special emphasis to some of the physical, psychologic and sociological factors" (JAMA 1942, 683). As one might expect, the review also drew out the medical implications, stressing "the increasing importance of engaging the medical profession, from the point of view of the practitioner and of the research investigator, for the alleviation of fatigue" (683). Eventually, the Fatigue Laboratory fell out of favor with the Harvard administrators, who in 1942 changed the name to the Laboratory of Industrial Physiology. From 1945 they allocated more resources to the nascent School of Public Health, and the laboratory was closed in 1947.

It is clear that the work of Mosso, Amar, and then Taylor and Gilbreth, led to roughly parallel transatlantic pathways during a crucial interwar period. The innovation of the application of physiology to industrial settings in the United Kingdom led to the Industrial Fatigue Research Board, and to the Harvard Fatigue Laboratory in the United States. In Britain, the IHRB reached its peak of efficacy through the 1937 Factories Act, which ruled on acceptable working hours, ambient conditions, and rest breaks. Along the way, encounters with unions and the need to recruit female members for the board meant that there was a politically progressive bent to this British research. The research of Taylor and the Gilbreths in the United States concentrated on maximizing factory efficiency, but again, the Harvard Laboratory applied extramural research to the measurement of human physiology in a range of environmental extremes. As I showed, this conjunction of physiological work on fatigue, and the need to investigate real-world implications during wartime for its definition and measurement, became an urgent scientific endeavor, and the resulting methods and charting of the progressive disruption of purely repetitive motor processes would lead to governmental recommendations and then policy on working hours, and eventually the rise of occupational health, by way of the science of ergonomics, human factors, and the emerging subdiscipline of industrial psychology.

There is, however, another pathway outside of this narrative of the subsumption of innovative physiological research into applied industrial contexts. Georges Canguilhem's influential 1947 lecture "Machine and Organism," for example, revisited the origins of the concept of a mechanics of movement, and therefore of the kinds of analogies after Borelli and Marey about biomechanics we saw in the previous chapter. Canguilhem traces the idea of "body-as-machine" to a 1696 text *Praxis Medica (The Practice of Physik)* by the physician Giorgio Baglivi, from the school Borelli founded, via René Descartes's well-known texts that explicitly equate humans to machines, such as his *Traité sur l'homme (Treatise of Man),* written before 1637 and only posthumously published, but ultimately to Aristotle's discussions on animal movement in *De Motu animalium (Movement of Animals)* and elsewhere. Of course, such analogies between human or nonhuman bodies and machines are familiar to scholars of the Early Modern period, as mechanical philosophy eventually displaced Aristotelianism. But whereas Aristotelian physics differentiated between natural movement and artificial movement of an organism or an object according to its purpose, or telos, such that a stone falling to the ground would be natural, but catapulting the stone would be artificial, Descartes's physics established an equivalence of motion. Hence the legacy of Helmholtz's *Arbeitskraft,* the transfer of forces across "animal" or "human" motors and machinery, as Canguilhem also recognizes. There are two points that contribute here. First, Canguilhem puts forward a counternarrative to this identification of the body-as-machine, of the machinic function of the human and animal organism. In fact, he considers the opposite view, that "the construction of machines can be understood by virtue of certain truly biological principles" ([1947] 1992, 45). Since there has been a current of anthropological thinking about tool use as extensions of human function in André

Leroi-Gourhan and others, Canguilhem reasons, "Machines can be considered as organs of the human species" (55). This is in fact a reversal of the Aristotelian legacy of the body-as-machine, since Aristotle had observed that the organs of animals were very specialized in their function. Canguilhem argues that the organism has a greater range of activity than the machine, is "less bound by purposiveness and more open to potentialities" (58). Thus, while parts in a machine are standardized and replaced, any given organ can perform a diversity of functions for the whole organism, and functional equivalences can occur through adaptation. In other words, assumptions at the heart of the rationalization of the "human-body-as-machine" (63) and the mechanization of the body in Taylor, sustained to varying degrees by the economies of energy of Helmholtz, and the measurement of industrial fatigue in Amar and Mosso, could be thought otherwise. The commonly accepted belief "that the generalized use of machines has slowly imposed contemporary industrialized society on man" (63) is not inevitable and, if machines are indeed organs of the human, implies that they can be far better adapted to the human body and its range of movements. Second, then, Amar's concept of the "human motor" not only relies on the rationalization of this false equivalence between bodies and machines, but the forms of power which animate them are categorically different. Canguilhem argues that "while a machine possesses motor power, it has no transformational energy that might propagate itself or be transmitted to an object outside the machine itself" (60). As with the idea of the standardization and replaceability of machinic parts, it is possible to append further mechanisms to a machine, to translate energy into lateral or rotational movement, say, or transmit movement from one part of the machine to another. But, unlike an organism, it cannot reorganize its parts into new arrangements, or create another machine out of itself. This point gets to the heart of Canguilhem's organicism, influenced as he is by Bergson and especially Goldstein (see chapter 2). The "motor" at the heart of the human or nonhuman organism would be fundamentally different in conception and articulation than Amar's troubling equivalency of the "human motor."

So far, we have traced an unusual and uneven alliance between the "new" science of physiology, with its interest and engagement with industrial production and worker fatigue, and the introduction of a series of graphic methods and instruments to measure this fatigue. We saw how the inaugural report of the Industrial Fatigue Research Board in 1920 was to examine the effects of repetitive activity and boredom on human subjects, the better to refine processes, correct errors, and reduce fatigue in the worker. There is certainly an understanding of the formation of motor habits within industrial contexts as almost entirely detrimental to the norms of productivity, error correction, and increased yields. There are other potential avenues for taking this research forward. As I discuss in the next chapter, during the same period that physiologists were transplanting themselves outside of the laboratory and grappling with the movements of bodies in industrial workplaces, another allied conceptual constellation was taking place, where neurological and philosophical lessons from brain injuries of soldiers were producing a new way of conceptualizing the phenomenological psychology of the body, and specifically the moving body. The following chapter therefore looks at the formation of Maurice Merleau-Ponty's concept of "motor intentionality," itself based on the neurological case studies of Aldhémar Gelb and Kurt Goldstein, among others. The kinds of cinematographic techniques that we have touched upon in terms of the physiology of fatigue will find kindred forms in the desire to get to grips with the very idea of *motricité,* an idea central to the spatiality of the body for Merleau-Ponty in his major philosophical work *Phenomenology of Perception* ([1945] 1992).

CHAPTER 6

Motricity

Merleau-Ponty, "Motor Habit," and the Neurology
of "Abstract" and "Concrete" Movement

Not all traumatic brain injuries have obvious or expected outcomes. Two case studies of cerebral lesions are illustrative. The first is the celebrated story of the railroad construction foreman Phineas P. Gage who ended up with a meter length iron rod lodged into his skull after an explosion in 1848. Initially there was neither motor impairment nor even loss of memory. In Gage's case, known to many students in psychology and neuroscience, despite extensive damage to his left frontal lobe, there was no loss of consciousness and the patient spoke normally after a few minutes. He was loaded onto a cart to see the doctor, but sat upright throughout the journey and recognized and greeted the doctor that evening, despite the fact that an ounce of his brain protruded from the hole at the top of his skull. Luckily the physician in question was John M. Harlow, who had extensive experience with horrific war wounds. Harlow communicated his initial findings two months later in the *Boston Medical and Surgical Journal*. Recounting the case and the procedures taken, he concluded that the case was not just of interest to the "practical surgeon," but was "exceedingly interesting to the enlightened physiologist and intellectual philosopher" (Harlow 1848, 393). The physiologist and the philosopher alike might justifiably wonder how Gage's speech and motor functions remained relatively untouched in the aftermath of such extensive brain tissue damage. But over time, more troubling and subtle aspects of Gage's behavior would emerge, and provide much for the physiologist and philosopher to debate. For the focus of the initial brief case study of 1848 was on the lack of damage to overall cognitive ability. However, after Gage's premature death in 1860, Harlow published a more detailed and controversial article that explained some major personality changes he summarized thus: "The equilibrium or balance, so to speak, between his intellectual faculties and animal propensities, seems to have been destroyed" (1868, 13). The past decades have revisited the case study extensively, and the common interpretation after Harlow, and advanced by Antonio Damasio (1994), that Gage descended into impulsive and often inappropriate behavior, is now contested. Years after the accident Gage was employed as a coach driver, and the psychologist Malcolm Macmillan looked into historical accounts to detail the "extremely complex sensory-motor and cognitive skills required" of that job (Macmillan 2008, 831; 2000, 104–6). The enduring popularity of Gage's case both then and now is primarily because it offered evidence both for and against the influential idea of the localization of brain function, first suggested by Paul Broca (1861).

The popularity of accounts of cerebral lesions in the decades following Gage's death led the noted neurologist David Ferrier to state that "cases of injury of one or other frontal lobe, without sensory or motor affection, are very numerous" (1878, 446). Although not so widely known, our second case study of cerebral lesion, and the one central to this chapter, occurred some sixty years after Gage's incident and was notable in the literature for providing more detailed observations about motor function and dysfunction. This case would also be extremely interesting for an intellectual philosopher, as Harlow had it, but rather than an enlightened physiologist this time there was an enlightened psychologist and also a neurologist. The psychologist was Aldhémar Gelb, the neurologist, Kurt Goldstein, and the philosopher Maurice Merleau-Ponty. In 1915, at the age of twenty-three, fragments from an exploded shell entered the skull of Johann Schneider, damaging his occipital lobe, an area responsible for vision processing. Like Gage, Schneider's wounds healed, and to an untrained observer he had the appearance of normal functioning, eventually reentering the workplace. However, unlike Gage, Schneider was placed under extensive observation in Goldstein's laboratory, and the unusualness of his case emerged. First, he showed difficulty recognizing objects and shapes without being able to "trace" the outlines of the objects with his eyes or hands, a condition called visual agnosia. Although capable of visually perceiving objects, he was effectively unable to grasp their meaning or significance. Second, if occupied in a task that involved motor skills, Schneider could not shift easily from one task to another. In fact, if he was asked simply to mime an everyday task such as brushing his teeth, he was unable to do so without going through the motions of all the actions that led up to the brushing motion. Unlike most peoples' everyday embodied experience,

Schneider had difficulty transposing routine movements or gestures from one motoric context or situation to another. From this, Gelb and Goldstein (1920) make a distinction between "abstract" and "concrete" movement, noting the tension between the forms of context-free repetitive movement that is regarded as "automatic" (e.g., miming the brushing of teeth), and the kinds of contextual movements based on a concrete "situation" that are accomplished unreflectingly by the embodied subject (e.g., the whole series of movements that precede and then follow the brushing of teeth). What fascinates Merleau-Ponty here is that Schneider exemplifies a dysfunction, a breakdown in our usual background processes of motor signification. Schneider's unusual inability to grasp perceptual situations as a whole also affects his ability to accomplish particular movements and motor tasks without guidance or external intervention, and is evidence of a lack of what Merleau-Ponty calls "motor intentionality" (1992, 127). Thus motor intentionality is revealed as a major constituent of our everyday unreflective interactions with the world. Gelb and Goldstein's initial case study of Schneider in 1920 led to a series of papers by Goldstein until the 1930s, and some scholarship in phenomenology and the philosophy of cognitive science has explored the historical significance of their work (e.g., Jensen 2009; Mooney 2011).

Given the overarching project of this book, Merleau-Ponty's take-up of the Schneider case fits neatly into the historical unfolding of the neurological and physiological discoveries in the late nineteenth and early twentieth centuries that pertain to the sense of one's own body, or as Merleau-Ponty terms it, *le corps propre,* first mentioned in his lengthy introduction to *Phenomenology of Perception* but which also features extensively in the first four chapters, where he outlines the limitations of mechanistic physiology and classical psychology in interpreting our lived spatiality (esp. 2012, 49–51). In highlighting the concept of motor intentionality at this juncture, I therefore revisit Schneider's landmark case study in the light of ideas already in circulation, and discussed in previous chapters, around movement and the industrial body. Consequently, we revisit ideas about the graphical forms of measurement made possible through scientific instrumentation in modernity, and the means of observing movement and establishing efficiency in industrial production. By filtering this through Merleau-Ponty's rather novel argument concerning motricity and motor intentionality, we trace how the uptake of a certain scientific literature informed the development of these ideas, and toward the end of the chapter gesture toward aesthetics and some of the innovative crossovers between the arts and the sciences of movement to better contextualize Merleau-Ponty's motoric contributions. In terms of the scientific literature, Merleau-Ponty's reading of Gelb and Goldstein's initial case study pitched "motor significance" and "motor intentionality" within background bodily processes as against the mechanistic explanations such as the "associative-reflexive" model prevalent at the time. Moreover, an aesthetic component of this rediscovery of motricity within physiology finds its analog through visual explorations of the mechanisms of organisms in motion in prior decades, especially the chronophotography of Eadweard Muybridge and Étienne-Jules Marey, of course (covered in chapter 4), some of the paintings of Georges Seurat and others, and the "physiological aesthetics" movement speared by Grant Allen and Charles Henry, all materializing in the 1880s.

Motility and the "Problem" of *Motricité*

The centrality of movement *(motricité)* in the lived spatiality of the body for Merleau-Ponty's *Phenomenology of Perception* is undeniable. One of the most important concepts in the neurological literature of the early twentieth century reemerges in his text: the "body schema" *(le schema corporel),* introduced more extensively in chapter 1. Its relevance here relates to motricity, as the body schema is premised on "the relation between spatiality and motricity," says Donald Landes (2012, xli), and leads to a specifically "motor intentionality" (112ff). This underlines the import of bodily movement for Merleau-Ponty's project as a whole. It is the third chapter of *Phenomenology of Perception* in particular where Merleau-Ponty focuses on the centrality of movement to embodied consciousness, at one point deeming *motricité* as "not the handmaid of consciousness" but rather the "motor grasping of a motor significance" (1992, 143). These thoughts on movement were later developed in his first course at the Collège de France (1953) in terms of the aesthetic encounter with movement and its traces, and expressive life. What he means by *motricité* is the focus of this section, and what he means by "motor significance" and "motor intentionality" is explained later when we revisit Schneider's case study.

There are two issues here about *motricité,* the word Merleau-Ponty picks up from physiological literature in the 1930s. First, there is the question of differences in translation of the term into English for both translations of *Phenomenology of Perception,* and why this word in particular has altered. The first English translation was by Colin Smith in 1962 (the 1992 version of this translation is cited here), and the subsequent secondary literature had *motricité* straightforwardly as "motility." The more recent translation by Landes in 2012 advances the more technical word "motricity," a term from physiology, in its place. The change in translation is offered with some justification, but both here and in his *Merleau-Ponty Dictionary* (2013), Landes provides a little etymological context. His rationale is that Merleau-Ponty uses the word *motricité* when discussing the Schneider case, which Landes claims is a physiological term in the translator's introduction: "The term indicates motor function, motor activity, and the power or faculty of movement," and Landes says these meanings are present in the English term "motricity," unlike the more common terms "motility" (used in the Smith translation) or "motivity" (Landes 2012, l). Landes's definition is mirrored almost exactly by the *Oxford English Dictionary,* where the origin of "motricity" in physiology is noted along with its definition: "Motor function; the faculty or power of movement by the body or a body part" (*OED* 2015b). Of course, its physiological origin is significant for Merleau-Ponty's uptake of the term. In writing *motricité,* Merleau-Ponty is actually transcribing the German term *Motorik,* a word which originates in A. A. Grünbaum's essay on aphasia, "Aphasie und Motorik" (1930), and that Merleau-Ponty cites several times, especially in part 1 of *Phenomenology of Perception.* In a footnote, Landes briefly touches upon how Merleau-Ponty's use of the term is transposed from physiology to his understanding of intentionality and Edmund Husserl's privileging of an embodied "I can" over an "I think" (Landes 2013, 133–34; see also Landes 2012, 511n7). The active nature of the English noun "motricity" over its more commonly used alternatives is stressed by Landes as a factor behind his choice, although the biological and medical uses of "motility," say, would indicate it is equally valid and even interchangeable. The definition of "motility" in the *Oxford English Dictionary,* for example, notes that the term comes from medicine and biology: "The power of active movement, esp. of a body part, cell, or organelle; the degree or type of such movement" ("motility," *OED* 2015a).

Second, there is the significance of the concept's origins in German physiology, and how Merleau-Ponty relates it to the lived experience of one's own body, *le corps propre*. The physiological concept of *motricité* will later be shown to have implications for the sciences and aesthetics of modernity, at a time when instruments of measurement could straightforwardly cross over from the realms of scientific observation and into art practice. Merleau-Ponty's understanding of the physiology of movement, by his own admission, was influenced by the research findings of Grünbaum, Goldstein, and Gelb, and the more speculative "sensory-motor process" Henri Bergson proposes in *Matière et Mémoire* (1896) that emphasizes "the unity of perception of action and events" and seeks to "involve consciousness in the world," according to a footnote in Merleau-Ponty's chapter on mechanistic physiology (1992, 91n19). In that same footnote, Merleau-Ponty summarizes Bergson's presupposition of an interior form of consciousness that separates the moving body from consciousness as such, since the body is present in space, but the mind stands outside the flow of time. For Merleau-Ponty, this has undesirable consequences: "But if feeling is representing a quality to oneself, and if movement is changing one's position in the objective world, then between sensation and movement, even taken in their nascent state, no *compromise* is possible, and they are distinct from each other" (91n19). Even if the body is part of a becoming of consciousness, thinks Merleau-Ponty, it remains an "objective" body. Such a conclusion is antithetical to his conception of forms of motor competence that arise within what he calls our "situation," and is a point of contention based as it is on the incommensurability of *le corps propre* and *le corps objectif*, the latter being the scientific–objective conception of the body in mechanistic physiology and classical psychology. The discussion of "situation" is particularly dense in part 1, chapter 2, and the following chapter of *Phenomenology of Perception*. As I go on to demonstrate, however, this distinction is not absolute, and discoveries in mechanistic physiology will spur more creative and even artistic engagements with human and animal bodies in motion outside philosophy and phenomenology.

By virtue of early neurological discoveries made in the first half of the nineteenth century, as we have seen, a medicalized language begins to emerge that categorizes and measures hitherto indistinct somatic sensations. In chapter 1 I established a narrative commencing with the so-called muscle sense of Charles Bell (*The Hand,* 1833), a comparable "muscular sense" and the coining of the term "kinaesthesia" from H. Charlton Bastian ("On the Muscular Sense and the Physiology of Thinking," 1869), the identification of sensations pertaining to movement became progressively legitimized through medical discourse, and a great deal of this arrived through nineteenth-century work on the so-called *Muskelsinn* (muscle sense) that was reported in English journals (see also Paterson 2013). But it is arguably in the late nineteenth and early twentieth centuries that the more radical possibilities of the sensorium and its capacity for alteration occurs as a result of the industrialization of the modern body. In the rapidity of change symptomatic of urban modernity, Walter Benjamin famously observed that "technology has subjected the human sensorium to a complex form of training" (2006, 191). Whether this technology is the profusion of traffic lights in major cities, leading to new forms of pedestrian awareness, or the repetitively staccato rhythms that shape the regularity of the factory worker's body through the conveyor belt, and that underlie the reception of that most democratically modern of art forms, film, Benjamin's point is rather general, although his examples remain potentially diffuse. Consequently, our focus has been on technologies that rendered hitherto uncharted internal bodily regions visible, that recorded the depths as legible surface by means of graphical representations (especially Marey's "graphic method" in chapter 4). A key aspect underlying Benjamin's training of the sensorium is the alteration of the rhythms of bodily movement according to industrial mechanisms on the factory floor, and therefore time is inevitably a factor in the measurement of that movement, and in the tracing and recording of the nervous impulses that stimulated and sustained such movements through graphical means. For, around the middle of the eighteenth century, nerves were categorically no longer the carriers of animal spirits but of electrical impulses, and therefore the nerve signal could be timed definitively. In 1867 Herman Helmholtz, the physicist and physiologist, recorded human nervous impulses as traveling between 35 and 45 meters per second. As Laura Salisbury and Andrew Shail summarize in terms of this period:

> Nerves were measured on, and understood in terms of, an axis—time—the representability of which had been established by the new speeds and frequencies achieved by modernity's pumps, springs, joists,

derricks, engines, factories, work crews, transportation networks and communication links. (2010, 30)

As a certain form of technological modernity became ascendant, in other words, new models of the body that engaged with such priorities of time and measurement would emerge. Furthermore, unlike previous models of the body that involved humors or animal spirits, the human body as a collection of measurable impulses was now deemed to be regulated by a set of finite energies—"limited energies" rather than "unlimited forces," as Salisbury and Shail put it (30). Once plotted through the new scientific tools of measurement, the dissipation of nervous impulses could now be graphically depicted.

Through increasing use of graphical instruments of observation and measure, then, the rise of such technologies transformed not only the forms of measurement but also the sensory capacities of the body. There is no shortage of historical analyses of the ways in which hidden movements and processes of the human body, determined through the touch or smell of a physician, say, were now rendered visible through technologies and so altered medicalized discourse (see, e.g., Foucault 2003; Borell 1993; Danius 2002). The skill of percussion is illustrative here, long taught to medical students since the eighteenth century, which involves placing the hands and fingers on the patient's body to determine the place, and the relative health, of organs through making percussive sounds and listening carefully to subtle differences (see, e.g., Nicholson 1993; Harris 2016). This hands-on practice was replaced with graphical representations of pulses, initially on paper through instruments such as the sphygmograph, an instrument invented by Étienne-Jules Marey for measuring and recording the pulse, and "considered a 'revolution' in medical technology" according to Anson Rabinbach (1992, 89). Other technologies furthered this revolution, including chronophotography, pioneered by Eadweard Muybridge in 1872 and also explored by Marey, the polygraph invented by John Augustus Larson in 1921, and Ferdinand Braun's cold-cathode diode of 1897, known as the Braun tube, a precursor to the cathode-ray oscilloscope in 1931. Such technologies enabled a universal and visual form of recording and reading of sensation, as opposed to the formerly tactile-acoustic form of percussion that, as a skill requiring prolonged practice, performed diagnoses through the carefully predetermined qualities of resonating bodies. On the one hand, this shift in the medical sensorium led to "a new kind of formalist observation in the sciences, in which complex dynamic processes were reduced to relatively simple visual and quantitative relations," summarizes Robert Brain (2008, 401), but this shift was not a straightforward retraining from tactile–acoustic to optical forms of observation. The larger significance was a shift in the relation between observable qualities. Namely, observable impulses that found graphical representation were not actually optical phenomena as much as what Peirce terms "indexicals"—that is, a "functional double of human functions," as Brain puts it, comparable to musical notation rather than direct observation (2008, 401). If listening and touching were a discrete or punctuated form of percussive measure premised on direct contact with living flesh, then the shift to graphical inscription offered a more continuous reading of bodily dynamics, a visual representation of changes in physiological processes as well as graphical snapshots of bodily status. Inevitably, this changed the relationship between the observer and the instrument observed. On the other hand, and what concerns us here, such indexical technologies of graphical representation paved the way for the means of recording other, less immediately given, forms of physiological phenomena. The surface of the skin, after all, could still be touched by the physician's hand and offer readings of vital signs. As we saw, the new trick was to produce forms of legibility of physiological processes far beneath the skin surface, or for some European physiologists studying evolutionary phenomena in experimental laboratories, to begin "to regard the animal body itself as a matrix of surfaces," in the words of Brain (402). In the case of early experimentalists such as Herman von Helmholtz and Wilhelm Wundt, human physiology was investigated as a path to scientific psychology, and graphical recording instruments might be considered as amplifications or "extensions of sense" (402). But in 1878 Marey rather contrarily argued that his new recording devices were acting "like new senses," and consequently could claim that this constituted "their own field of investigation" (Marey 1878, xiii).

Given this new potential realm of concretely documenting and interpreting what had hitherto been unexamined, of making physiology into a legible matrix of surfaces, the question of animal and human motricity that so intoxicated Marey becomes highlighted (see chapter 4). The production of the series of graphical images that captured human bodies in movement, such as with Marey's *fusil photographique* of 1882, the rifle-like photographic device that took a series of twelve sequential photographs, made concrete movements legible in a way that no other technology had managed, and therefore a whole new field of investigation had indeed opened up. Marey's physiological investigations were startlingly ambitious, charting locomotion across the zoological spectrum, from protozoa and primitive marine life right up to galloping horses and the human gait, in order to understand the patterns behind evolving physiological rhythms. According to Brain, such laws or patterns of movement would be understood in relation to energy expenditure and morphological evolution, no less (2008, 405). With the engineer Charles Fremont, Marey extended his zoological observations to study the locomotor physiology of contemporary artisans, producing the now iconic chronophotograpic images of the motions of modern labor. Consequently, the new science of the observation of the stages of human and animal movement was coextensive with the rise of the observation of human bodies as motor potential and, in Marey's petitions for funding to the French government in the 1870s and 1880s, he made a virtue of the applications of these techniques for the conservation of labor power and to enhance military training (Rabinbach 1992, 104). Fremont's intention in observing the movements of modern artisans was to improve labor practices but also, more ominously, to develop automated procedures for modern production methods. As Brain puts it, "Early *automation* would here follow from the proper understanding of the evolution of *automatisms*" (2008, 405; original emphasis). As we saw in chapter 5, observations of human labor were to become even more rigorous and formalized through methods of what F. W. Taylor termed "scientific management," what became known pejoratively as "Taylorism" in Europe and the United States from around 1913. Within scientific management, methods such as filming the movement of workers was developed as a distinct methodology by Frank and Lillian Gilbreth as time and motion studies. The physiological impetus to observe movement through the matrix of surfaces of both animals and men, aided by photographic, chronophotographic, and filmic means, arguably starts as a grand comparative biological survey, yet also results in the close observation of small-scale repetitive motions and bodily skills, the better to streamline industrial processes and maximize productivity. We now examine some related and contemporaneous research on reflexes, habit, and automatism in physiology and aesthetics before returning to

Merleau-Ponty's concept of motor habit and motor intentionality in the final section.

Automatism, Physiology, and Aesthetics

A link between automatism in the nervous system and automatism in aesthetics will be pursued below through an example of the Post-impressionist painting of Seurat, since any investigation of the painterly depiction of movement of bodily forms during that period inevitably gestures toward the contemporaneous scientific approach of the chronophotographers. Alongside any such alliance between modernism and the physiological underpinnings of the perception of movement, there is also the matter of the historical relationship between neurology and modernity, which has been acknowledged only intermittently. Jonathan Crary specifically identifies the role of neurology in generating discourses of modernity, for example. He claims that "a vast range of modernizing projects in the West well into the twentieth century, and more specifically a new inventory of the capacities and functions of an attentive observer" (2001, 164) would simply not have been achieved without the emerging consensus, widely held by the 1880s, that the human nervous system operated as a system of reflex functions distributed throughout the body. Hitherto, since Thomas Willis's *Cerebri anatome* of 1664, it was assumed that the brain was central in the organization of nervous communication. Willis was in fact the first to coin "neurology," as a Greek word in his Latin text, as "a Doctrine of the Nerves" (see, e.g., Feindel 1962). On the other hand, the observation of the automatic nature of animal patterns of movement without conscious thought or intention, or "automatism," meant that reflexes and motor habits were of early interest to neurology. Charles Sherrington often gets the credit for discussing the "reflex arc"—that is, the neural pathway that controls an action reflex in an animal through the motor neurons in the spinal cord, bypassing any direct relay to the brain (see chapter 1). However, from 1832 the English physiologist Marshall Hall had begun to publish influential studies on the reflex arc that affected theories of nervous communication (Hall 1833). The types of experiments he performed were by now quite familiar and nothing new, observing apparently voluntary actions by decapitated amphibians in response to stimuli. From these, Hall built a theory of "an entire catalogue of reflex arcs, linking sensory and motor nerves distributed to all points in the body, that was built solely into the spinal cord, and which, by sensing both external and internal changes, determined a vast range of bodily processes" (Salisbury and Shail 2010, 18). Hall found that automatic and nonconscious processes that did not require active participation or control by the brain, such as breathing, could still be taken over and consciously influenced when the will determined it. Conversely, when reflex arcs were activated, the brain's awareness of them was superfluous, entirely unnecessary for the reflex action to take place.

The related doctrine of "automatism" soon became a recognized area of inquiry in mid-Victorian England, fascinating such eminent biologists as T. H. Huxley, Herbert Spencer, William Benjamin Carpenter, and Henry Maudsley. Carpenter, a physiologist and zoologist, had written essays including "The Automatic Execution of Voluntary Movements" in 1850, and "On the Doctrine of Human Automatism" in 1875, and as early as 1852 noted that "reflex actions" were universally recognized by physiologists, a phenomenon where "impressions made upon the nervous system are followed by respondent automatic movements" (Carpenter 1888, 169). Carpenter's collection of essays *Nature and Man* (1888) further assumed a decentralization of the brain as the principle organizing unit of reaction and motor response for an organism was underway. As part of a discussion on the same types of experiments that Hall and others had previously conducted—namely, exciting a muscular contraction in an animal by galvanizing the motor nerve of a limb—Carpenter similarly noted that electrical transmission does not go directly to the part of the brain responsible for sensation, what he calls the "sensorium," but rather manifests as a "nerve-force" that excites the part of the body directly preceding the limb. Just like Hall, Carpenter presumed the same principle of automatism observed in muscular contraction is to be found in other organismic processes like nutrition and secretion. Carpenter considered these processes as manifestations of what he calls a "general organizing force," being "the nervous substance giving up its characteristic organization whilst developing nerve-force, and that nerve-force being transmitted to a distant part, to be applied there in producing or modifying organization" (181). In other words, he had discovered the principle that the quantity of organization lost centrally is equal to the quantity of organization gained at the periphery. If the brain's awareness of the stimuli activating reflex arcs was not always necessary for triggering reflex actions, this gave a newfound autonomy to the spinal cord, and suggested that reflex actions might still occur despite damage to the nerves of sensation. In Hall's *Memoirs* (1837), the preconscious nature of these reflexes was termed the "excito-motory" property of the spinal cord, which presided over muscular activities all over the body, keeping sets of antagonistic muscles in static tension, for example. Indicating the ongoing and dispersed nature of these muscular actions, and also the lack of conscious awareness or control, Hall noted: "The reflex function exists as a continuous muscular action, as a power presiding over organs not actually in a state of motion" (in Salisbury and Shail 2010, 18). The discovery of such excito-motory properties indicated the existence of a whole new intermediate nervous system, separate from the conscious or volitional nervous system, yet remaining distinct in turn from the vegetative nervous system common between higher and lower animals. It was distinct because of the dynamic nature of the constant flux of external and internal stimuli bypassing conscious awareness or sensation. Looking back on such findings toward the end of the century, Carpenter similarly distinguishes between automatic movements of two kinds: first, the "excito-motor" effected purely through the instrumentality of the spinal cord and associated nerves, without any felt impressions; and second, the "sensorimotor," "in which sensation necessarily participates, the respondent motions not being executed unless the impressions are felt," which occurs through "the chain of sensory ganglia" between spinal cord and cerebellum (1888, 169).

In the nineteenth century, observations of automatism within physiology and psychology such as these had their correlate within other disciplines, including the arts. For example, the philosopher Stanley Cavell used the term "automatism" in this context to describe the forms, techniques, and conventions "that arise creatively out of the materials and conditions of artistic practice" and as potential forms or conditions for future practices, and that "habits, codes and materials set the limit-conditions for subjectivity, creative agency, and artistic process," in the words of Brain (2008, 399n22). Although space does not permit extensive exploration of examples of automatism in art, the broader definition of automatism offered by Brain is useful, as he returns to the original physiological sense in order to make it more generally applicable across disciplinary areas, with its "specification of the automatic, self-acting, and machine-like character of many organic movements and psychological actions which occur without the direction of the mind or will" (399n22). Seurat's paintings, for example, are regarded by Brain as reflecting the multiple meanings of automatism in fin-de-siècle culture. Indeed, the pointillist depiction of dancing bodies in *La Chahut* (1890) partly resembled Marey's high-speed photographic studies of moving bodies, with the sense of a moment in movement caught through mechanized perception. The dancers, with their developed and refined motile competences, would at least partly involve a decentered automatism without the conscious direction of the mind toward their limbs. Published the same year as *La Chahut*, the chapter on "Habit" of William James's *Principles of Psychology* furthers this formulation of automatism as the offloading of tasks performed by consciousness, to make the nervous system "our ally": *"For this we must make automatic and habitual, as early as possible, as many useful actions as we can,"* and so the more "details of our daily life we can hand over to the effortless custody of automatism, the more our higher powers of mind will be set free for their proper work" (1890, 122, emphasis in original). The possibility of emancipation through the production of new automatisms acknowledges a creativity in habit-formations, something common to movement but also artistic practices. The question of automatism therefore went deeper than the self-acting mechanisms involved in recording and photography, reflected in Marey and Seurat. Might artistic practices themselves be considered as automatisms in this case, wonders Brain, "etched deeply into the craft and possibly even the physiology of artists and artisans" (2008, 408)? Furthermore, the co-arising of automatism and modernity in a rather specialized area of modernist aesthetics that aimed to interrogate the physiological processes behind aesthetic experience, known as "physiological aesthetics," would not be accidental, and would involve measurable automatic responses by the body. Grant Allen's book of that name, published in 1877, was explicitly indebted to Helmholtz's work, including *Handbuch der Physiologischen Optik* (1867), and dedicated to Herbert Spencer, the biologist of automatism and the originator of the related concept of "aestho-physiology" which was first mentioned in the second edition of his *Principles of Sociology* (Spencer 1872, 533–40; Allen 1877). Whereas Spencer's "aestho-physiology" sought to explain emotions and conscious states in terms of "simulations and discharges of nerve-force reverberating throughout the nervous system," in Crary's words (2001, 171), Allen's project of physiological aesthetics attempted to put those most subjective sensations of pleasure and pain that arise when encountering an artwork on a more scientific and measurable footing, providing what he notably terms a "diagrammatic account of physical and physiological facts" within aesthetic encounters (Allen 1877, ix). The diagramming of the sensorium, once again, was continuing the process of turning physiology into a legible surface, just as Marey was doing elsewhere. Alongside Allen, the scientist and aesthetician Charles Henry also popularized the concept of physiological aesthetics from 1885 onward, which was to have a significant impact on both Seurat and Paul Signac (Henry 1885). For Johanna Drucker, Seurat's work exhibits two features of modernism in particular that are relevant here: first, the "scientificization of technique," or at least a rational underpinning for the depiction of movement in painting; and second, his consideration of the canvas as "an autonomous and abstract space" that might conceivably be regarded as analogous to the screens of more scientific forms of graphical inscription (Drucker 1996, 41). Whereas bodily movements in themselves seemed endlessly fugitive, Henry described how the new graphic recording techniques could open up such movements to novel forms of analysis:

> One could not study the subjective impression of movements, of colours, of musical and articulated sounds when natural philosophy had not yet distinguished, classified, and formulated its objects, when nothing was known about the composition of forces, the decomposition of the spectrum, the nature of musical intervals, of consonants and vowels, when the physiology of the nerves was not yet anticipated, when the sense of the evolution of language was absolutely lacking. The scientific culture of recent centuries and above all of the present has overcome these obstacles. It seems that the problem is ripe. (Henry, in Brain 2008, 405)

The continuation of the decentralization of the brain was furthered by Sherrington in *The Integrative Action of the Nervous System* of 1906, where he conceived of the disparate organs and components of the body as a series of "semi-autonomous private units" (Salisbury and Shail 2010, 30) that effected functional coherence through the activity of the nervous system. In introducing this idea of the integrative nature of the nervous system, Sherrington explains that "in the multicellular animal, especially for those higher reactions which constitute its behavior as a social unit in the natural economy, it is nervous reaction *par excellence* which integrates it, welds it together from its components, and constitutes it from a mere collection of organs [as] an animal individual" (Sherrington 1906, 2). Incidentally, some metaphorical implications of the integration of semiautonomous units into a functional whole, and the overall tendency of decentralization, can be observed in other realms. In their introductory essay to *Neurology and Modernity,* for example, Salisbury and Shail (2010) offer the analogy between this semiautonomous picture of the nervous system and the contemporaneous organization of the political economy of the state, since during this period the state increasingly came to function as a noncentralized economic network. Elsewhere, Norbert Wiener's vision for cybernetics involved an awareness of reflexes and the reflex arc within the human operator of his anti-aircraft prediction system, as Peter Galison (1994) shows.

From the conception of a decentralized chain of reflexes and automatisms in *le corps objectif* developed into the early twentieth century, we now turn to the kinds of experiences of motricity of *le corps propre* during this time. Between 1906 and 1935, especially, neurologists were forming ideas that concerned the double function of sensations involved in the sense of having and moving one's own body—that is, "proprioception" (for Sherrington 1906), "body schema" (for Head and Holmes 1911), and "body image" (for the psychologist Schilder in 1935, as we saw). Sherrington had coined "proprio-ception" in *The Integrative Action of the Nervous System* to articulate the body's awareness of itself, its movements, and position in space. Likewise, the phrase "body schema" was first described by Head and Holmes in a paper on "Sensory Disturbances from Cerebral Lesions," and developed further in Head's book *Studies in Neurology* in 1920. Although the basic concept was discussed in this book's introduction, we briefly revisit it in order to pursue some neurological implications for the case study in the next section. The body schema is "a system of pre-conscious, subpersonal processes that play a dynamic role in governing posture and movement," according to Satz (2010, 123). Satz considers this proprioceptive body to be some kind of somatosensory "phantom body" that inhabits, and for the most part coincides with, our phenomenological body, and makes the analogy with Silas Weir Mitchell's discovery of the phantom limb phenomenon. Mitchell first wrote of this in an article from 1871, and explored it further in his 1872 book *Injuries of Nerves and Their Consequences*. A wealth of such neurological injuries and unmapped symptoms were discovered and archived as a result of the American Civil War, and Mitchell describes cases where the nervous system is out of sync with the rest of the body, pain is mismapped, and parts of the body seem to be telescoped or metamorphized. In the final section of his book, "Neural Maladies of Stumps," Mitchell relies on descriptions of the subjectively felt experience of pain and sensation to focus on "sensory hallucinations" ([1872] 1965, 348ff), where the phantom limb phenomenon appears once again. "Nearly every man who loses a limb carries about with him a constant or inconstant phantom of the missing member," observes Mitchell, "a sensory ghost of that much of himself . . . faintly felt at times, but ready to be called up to his perception by a blow, a touch, or a change of wind" (348). Rather remarkably, Mitchell was able to induce phantom sensations of missing limbs through applying an electrical current to nerves in the severed stump, a process referred to as "faradising" (349). But it is as a result of the profusion of neurological dysfunction that occurs in war injuries that a precursor of the proprioceptive body was stumbled upon and described. In terms that preempt both Sherrington's "proprioception" and Head and Holmes's concept of "body schema" by decades, Mitchell writes:

> We are very competent in health, even with closed eyes, to know where and how far removed the hand may be at any moment, and this knowledge is the result of long-continued and complicated sensory impressions, ocular, muscular, and tactile. Should we lose these by amputation, we cease to have consciousness of the extremity of a limb as set at any fixed distance. (351)

Head and Schilder also refer to the phantom limb phenomena that Mitchell had identified, and considered it an example of body-schema distortion. Although Schilder referred to Head and Holmes's account of "body schema" and "postural schema," he instead preferred the term "body image" in his book of 1935, since it also had connotations of a phantom "other" body that remains somehow related to our own body as we experience it: "The image of the human body means the picture of our own body which we form in our mind, that is to say the way in which the body appears to ourselves" (Schilder [1935] 2013, 11). Head and Holmes's body schema stressed the dynamic nature of this set of processes relating to bodily posture and movement, and the word "schema" denotes a kind of "standard" or postural template of the current bodily position in space, a motoric and visual phantom as it were, against which any alterations or further movements are judged:

> By means of perpetual alterations in position we are always building up a postural model of ourselves which constantly changes. Every new posture or movement is recorded on this plastic schema, and the activity of the cortex brings every fresh group of sensations evoked by altered posture into relation with it. Immediate postural recognition follows as soon as the relation is complete. (Head and Holmes 1911, 187)

It is immediately after this point in their essay that Head and Holmes discuss the perturbations of this postural and positional conception of one's own body that occur when a limb has been amputated, and therefore directly invoke the phantom limb phenomenon. Merleau-Ponty's discussion of phantom limb syndrome is consequently part of his project of rethinking the spatiality and motricity of the body, and at one stage his use of "body schema" directly echoes that of Head and Holmes: "I hold my body as an indivisible possession and I know the position of each of my limbs through a *body schema*" (2012, 100–101, emphasis in original). What is specific to Merleau-Ponty's formulation is that, even if motricity is experienced by the subject as an "I can" of bodily possibilities, it is also an originary intentionality that takes account of habits and acquired motoric patterns, where "our field of experience is structured according to a tacit set of sedimentations and possibilities," as Landes (2012, xl) puts it. Such sedimented motor habits structure the perceived world of the subject and consequently offer a new configuration of possible actions. Now, in the same way that phantom limb phenomena famously highlight what the "normal" functioning of a body schema consists in, including the sedimentations of motoric actions and anticipations due to the "habitual body," a sensory-motor dysfunction in another neurological case study will helpfully "clarify the original intentionality of motricity in normal experience" for Merleau-Ponty, thinks Landes (xli).

"Concrete" and "Abstract" Movements: The Case of Schneider

No doubt some readers are already familiar with the celebrated neuro-logical case study of Schneider, a patient who suffered brain injury in the First World War. Initially known as "Sch" in the psychology literature for the usual reasons of anonymity, Schneider was wounded by shell fragments in 1915 at the age of twenty-three. The fragments penetrated his skull and occipital lobe, an area associated with visual processing. Outwardly he appeared to be healing, exhibiting no immediate sensory deficits, and it was only after extensive testing in Goldstein and Gelb's laboratory that they found Schneider had visual agnosia, difficulty in rec-ognizing objects and shapes, and was therefore psychologically blind. Unbeknownst to him, after his wounding Schneider had developed a se-ries of adaptive strategies to pursue everyday activities, including read-ing alphanumeric figures through tracing small head and hand move-ments, along with adaptations for his "figural blindness" so that he could determine the difference between ordinarily different categories of objects such as humans and vehicles (Pickren 2003, 129). Schneider's case study features even in current psychology textbooks, but the sig-nificance of Gelb and Goldstein's initial paper of 1920 was that it was written in response to the accepted view, after Goldstein's tutor Carl Wernicke and Paul Broca, of the localization of cerebral function (see, e.g., Broca 1861). Essentially, this orthodoxy held that brain functions operated in isolation, and consequently explanations for complex ac-tivities such as reading relied on the stimulation of discrete brain areas that were rather reductively associated with particular functions, with Wernicke further developing this into a system of reflexes known as an "associative-reflexive" model of brain function (130). In their exten-sive laboratory interaction with Schneider, Gelb and Goldstein found that this model was inadequate to the task of explaining how isolated brain functions allow individuals to grasp situations in all the richness of their natural context—that is, to produce a gestalt. The perceiving subject is demonstrably *not* a series of discrete brain functions that link up through ordered chains of association but, through the reception of sensory stimuli, the subject grasps the significance of a contextually rich situation in order that tasks or goals may be accomplished.

This is why Gelb and Goldstein's 1920 paper forms a major empirical strand in Merleau-Ponty's chapter on motricity and the body. For, it will

turn out, Merleau-Ponty follows Gelb and Goldstein's distinction between "abstract" and "concrete" movement, noting the tension between
the forms of context-free repetitive movement that we have previously
determined as "automatic," and the kinds of contextual movements
based on a concrete "situation" that are accomplished unreflectingly
by the embodied subject. In relation to the Schneider case, Merleau-
Ponty's analysis privileges those naturalistically flowing or "unbroken"
movements usually accomplished in what Kelly terms "skillful, unreflective bodily activity" (2002, 377) exactly because our experiences of
movement are never limited to habitual actions, or effectively "set and
automated responses to situations of a certain type," thinks Timothy
Mooney (2011, 360). Ordinarily, that is, in unreflective bodily activity,
"my body appears to me as an attitude directed towards a certain existing or possible task. And indeed its spatiality is not . . . *a spatiality
of position,* but a *spatiality of situation,*" claims Merleau-Ponty (1992,
100, emphasis in original). What the philosopher terms "motor intentionality" therefore denotes those intentional activities that essentially
involve our bodily, situational understanding of space and spatial situations, and involves voluntary motor activities such as reaching, grasping, and pointing. Each of these types of prehension is directed at a
particular object at a particular location, "not as an object represented,
but as that *highly specific* thing towards which we project ourselves," as
Merleau-Ponty puts it (138, emphasis in original), and involves a particular configuration of anticipatory movements with that object in mind.
Different objects such as a coffee cup on a desk, a door knob, or a type-
writer keyboard offer their own "motor signification," says Merleau-
Ponty—that is, their own possibilities for use and interaction (126–27).
"When I grab for my coffee cup in the morning I direct my activity toward it, not simply toward some independent location that it occupies,"
summarizes Kelly; "Genuine grasping, it seems, is directed not just toward a location, but toward a located object" (2002, 384). In fact, as a
result of Gelb and Goldstein's paper, and due to more recent experimental evidence considered below, the directedness toward an actual object rather than an imagined object will become significant for exactly
this reason. To develop this argument about motor intentionality in the
light of Merleau-Ponty's reading of Gelb and Goldstein's explanations of
the gestalt, we now start to build upon the historical background with

its explanations for motor phenomena in terms of reflex, automaticity, and motor habit.

Even with repetitive motor tasks involving actual objects, such as working in a factory, what Merleau-Ponty terms "motor significance" still pertains. This type of significance requires that habitual actions are meaningful and not mere reflex movements, while at the same time suggesting it is not the result of mental representations: "The cultivation of a habit is indeed the grasping of a significance, but it is the motor grasping of a motor significance," he explains (1992, 143). He then offers immediately recognizable examples such as a woman with a tall feather in her hat negotiating her way through doorways, a blind man's use of his stick, and an experienced driver's ability to negotiate small parking spaces. It is significant that all these examples are based not solely on repetition or reflex as such, as learning a new dance move, or getting up to speed with the manipulation of an object on a factory conveyor belt would emphasize. Rather, Merleau-Ponty's examples are designed to demonstrate the acquisition of a set of motoric skills that naturally become part of an expanding repertoire within the body schema. As Merleau-Ponty summarizes: "To get used to a hat, a car, or a stick, is to be transplanted into them, or conversely, to incorporate them into the bulk of our own bodies" (143). The extension of the body schema to objects outside the body, and even the specific example of a feather in a woman's hat, is lifted directly from Head and Holmes's 1911 paper, influential upon, but not directly referencing, the gestaltists:

> It is to the existence of these "schemata" that we owe the power of projecting our recognition of posture, movement and locality beyond the limits of our own bodies to the end of some instrument held in the hand. Without them we could not probe with a stick, nor use a spoon unless our eyes were fixed upon the plate. Anything which participates in the conscious movement of our bodies is added to the model of ourselves and becomes part of these schemata: a woman's power of localization may extend to the feather in her hat. (1911, 188)

Of course, the Schneider case is where this behavior, and the idea of the subject as essentially a motorically coping body within a spatial situation, starts to break down. It is important for Merleau-Ponty that the varieties of motor intentionality for "normal" functioning involve a certain plasticity that encompasses adaptive and creative, as well as habitual and repetitive, movements. Of the latter, in the kinds of repetitive movement required of a factory worker, or even the haptically engaged repetitive skills involved in hand-stitching wallets, which in fact became Schneider's job after recovering from injury, there is at least the possibility of transposing or adapting habits acquired in one "concrete" situation (say, a factory producing object X) to a different but comparable "concrete" situation (producing object Y). Gelb and Goldstein, and later Merleau-Ponty, agree that Schneider has difficulties with this transposition, ascribing to him an inability to re-create actions or mimic skills that are devoid of context for him. Asking him to act out everyday motor skills such as brushing teeth or shaving would involve having to mime a series of unrelated actions, all in order to reestablish his own motor context. This is regarded as an inability to produce "abstract" rather than "concrete" movements. In another example, this time from Merleau-Ponty's *Phenomenology of Perception,* ordinarily a typist can transpose their ability to type on one keyboard to another, a bodily skill which is neither premised on knowledge of the placement of letters on keys, nor on a "conditioned reflex" for each key, or "involuntary action." Instead, he continues, it is an instantiation of "knowledge in the hands, which is forthcoming only when bodily effort is made, and cannot be formulated in detachment from that effort" (1992, 144). Thus, motor intentionality involves a way of being directed toward the particularity of objects, and a bodily anticipation toward them. Speaking initially in this passage of Schneider, Merleau-Ponty first defines the concept, helpfully articulated in terms that lie between the movement involved in *le corps objectif* and that of *le corps propre:*

> What [Schneider] lacks is neither motility nor thought, and we are brought to the recognition of something between movement as a third person process and thought as a representation of movement— something which is an anticipation of, or arrival at, the objective and is ensured by the body itself as a motor power, a "motor project" *(Bewegungsentwürf),* a "motor intentionality." (110)

What happens if there is an obstruction in the process, or a modification of an action is required on the production line or in stitching wallets?

Then bodily anticipation breaks down. According to Mooney, Merleau-Ponty overemphasizes the rigidity of habit in Schneider in his reactions to situations, characterizing them as "automatised responses" to situations, a "rigid automatisation of the concrete" that can be associated with those mechanized models of classical psychology (2011, 360). But, interestingly, Merleau-Ponty draws from a different nonmechanistic tradition of thought on habit including Maine de Biran and Félix Ravaisson to think about a concretion of habit. In learning a new dance, adding to the existing repertoire of movements within the body schema, there is an alteration of "general motricity" in Landes's translation, but for this to take place the body must have received a "motor consecration"—"the body . . . 'catches' *(kapiert)* and 'understands' the movement," hence the motor grasping of a motor signification (2012, 144). The adaptation required in transposing movements from keyboard to keyboard, or for an organist from one organ to another, involves an alteration of our motor habits when such a "motor space" stretches out before our hands, a situation that does not offer a "spatial trajectory" for our fingers as much as a "certain modulation of motricity" (145). In keeping with previous physiological concepts about movement and automatism, the potential for such modulations is present even in the most preplanned concrete movements of Schneider, as his adaptability within a postinjury rehabilitative context was so rapid.

By bridging previous neurophysiological discoveries with the development of Merleau-Ponty's phenomenology concerning the experiences of the motricity of the body, we have made preliminary investigations of aesthetic applications of this productive scientific and philosophical nexus of ideas. Previous chapters have already raised the issue of the "body schema," the means of measuring bodily sensations such as pain and fatigue through scientific instruments, and the visual means of charting the human body in motion. But the theme of this chapter was the place of motricity in the lived spatiality of the body. As suggested in the section above on automatism, there are other rich potential avenues of research on the aesthetics of the moving body during that historical period. For example, from the second half of the nineteenth until the mid-twentieth centuries, Modernist choreographic experiments with muscular movement and expression, including the various forms of graphical notation that developed alongside what François

Delsarte termed "applied aesthetics," and later the eurhythmics of Émile Jacques-Dalcroze and dance-movement techniques of Rudolph Laban. More recent fascinations with so-called kinaesthetic empathy, and the theorization of the intersubjective nature of an embodied spectator watching dancers onstage, has been of interest in dance studies for a while, but has begun to be discussed in philosophical aesthetics. The dancer–philosopher Barbara Montero, for example, neatly frames this issue in terms of the problematics of "proprioceiving a dancer's movements" (2006, 238), where the audience does not simply perceive the movement of a body on stage from a disembodied perspective, but perceives the movement of another through their own body. Alternatively, in the realm of industrial production, we saw in the previous chapter how the inaugural report of the Industrial Fatigue Research Board in 1920 chaired by Charles Sherrington was to examine the effects of repetitive activity and boredom on human subjects, the better to refine processes, correct errors, and reduce fatigue in the worker. Furthermore, Schneider's return to repetitive labor, the stitching of wallets, shows a particular ideal of the industrial worker on the production line. There is certainly an understanding of the formation of motor habits within industrial contexts as almost entirely detrimental to the norms of productivity, error correction, and increased yields. As I hope to have shown, however, the plasticity, adaptability, and the powers of transposition of movements across spatial contexts point to more creative aspects of movement for human subjects. Merleau-Ponty's concern with the spatiality and motricity of the body can therefore be regarded in relation to a series of interests in the arts and sciences with movement, the physiology of motricity, and various means of representing and diagramming the movements of the human body. Ultimately, motricity, that is, the physiological capacity for the movement of the organism as such, is indissociable from the lived spatiality of the body.

ACKNOWLEDGMENTS

My first real engagement with sensations of the moving body and the idea of the "muscle sense" was an invitation to the "Touching and to Be Touched—Kinesthesia and Empathy in Dance" workshop at the University of Berlin in 2011, and in 2013 an invited paper for the Einstein Forum in Potsdam, Germany, organized by Dominic Bonfiglio. From 2013 onward I discovered the Society for Literature, Science, and the Arts (SLSA), whose conferences have been unexpectedly instrumental as a venue for raising ideas and generating feedback. Papers presented at its conferences in Notre Dame, Atlanta, and Toronto were indispensable pieces in the complex puzzle that became the first chapter.

An early version of chapter 2 on pain was stimulated by an invitation to give a paper on the occasion of the thirtieth anniversary of the publication of Elaine Scarry's seminal *The Body in Pain* at the University of Brighton, UK, in December 2015. Unbeknownst to the organizers, Leila Dawney and Tim Huzar, writing my paper prompted a radical thinking of the scope of this book and resulted in an article in *Body and Society.* Thanks also to Charles T. Wolfe for inviting me to the Organicism Workshop at the Department of Philosophy and Moral Sciences, Ghent University, Belgium, where my paper "On Pain from the Organismic Perspective: Goldstein *contra* Sherrington" helped the transition from journal article to book chapter.

Ideas about the oculomotor system that became chapter 3 came together rapidly and led to a series of talks in several architecture departments. First, thanks to Seher Erdoğan Ford for inviting me to the Department of Architecture at Temple University in 2017, to Athina Papadopoulou and Terry Knight for inviting me to the MIT Department of Architecture for its Spring 2018 Lecture Series, and to Jonathan Hale in the Department of Architecture of the University of Nottingham, UK, for inviting me to give an International Guest Lecture sponsored by the Sensory Studies Research Network and Centre for Critical Theory in 2018. These papers helped road test and refine my approach to architectural theory and archaeological examples.

The nascent ideas on fatigue for chapter 5 were first presented at the American Sociological Association (ASA) Section on Science, Knowledge, and Technology in Montréal in 2017, and then a more mature version at the International Sociological Association (ISA) meeting in Toronto in 2018. But by far the most comprehensive and useful feedback anyone could wish for came from the University of Pittsburgh Humanities Center Colloquium earlier in January 2018. My paper was an advanced draft of the version that appears in this book, and I am supremely grateful for the generous and helpful critical engagement that the respondents from the University of Pittsburgh, Mari Webel (Department of History) and Tomas Matza (Department of Anthropology), provided, as well as other attendees.

Indeed, the University of Pittsburgh has offered a wealth of resources and support over the years. In 2015 Mazviita Chirimuuta and I were recipients of the Special Initiative to Promote Scholarly Activity in the Humanities for our project "Evolving Concepts of Body Sensation and Motor Control in the Neuroscience of Movement," which allowed eye-opening archival research in Montréal and London. I will be forever grateful for my Dietrich School of Arts and Sciences Humanities Center Fellowship in 2017–18 for the teaching relief, of course, but also the community of supportive scholars. I especially wish to thank the former director, Jonathan Arac, for his generosity of spirit and critical perspectives. My colleagues in the Department of Sociology at the University of Pittsburgh have been supportive, and in particular I thank Professor Lisa Brush for reading chapters and making extensive comments in her usual thoughtful and rigorous way.

David Howes from Concordia University Montréal has been an unofficial mentor from my graduate studies onward, and his tireless advocation of sensory studies for decades has helped establish the importance of the senses in cross-disciplinary dialogs around the world. Both he and Tom Sparrow (Slippery Rock University) have offered helpful and incisive comments on this book at various stages. As an intellectual fellow traveler, I also thank David Parisi of the College of Charleston for long, fascinating, late-night discussions when our trajectories regularly converged at conferences, at panels that we co-organized, and in personal travels.

Last, and far from least, my wife, Mazviita, has been a stalwart, an intellectual sparring partner, reading and commenting from an allied but different disciplinary perspective. This book is dedicated to our constantly moving darlings Claire Tinashe and Lucius Tendai.

NOTES
- - - - - - -

1. The "Muscle Sense" and the Motor Cortex

1. The ancient text *Problemata* (Problems) is not to be confused with the much later text known as *Problemata Aristotelis*, which circulated around Europe in the Early Modern period. See, e.g., Blair 1999.

2. As Finger (2005, 159; 2001, 41n8) and others have noted, the story of the 1881 Physiology Society meeting took a dramatic turn when, three months later, Ferrier was summoned to a police station to answer charges filed by the Victoria Street Society for the Protection of Animals from Vivisection. This became a cause célèbre for the animal rights movement, as it was reported around the world. The charges related to operating on animals without a government license and in a way that caused them pain. Ferrier got off on a technicality, as in fact the surgeries had been performed under general anesthetic by Yeo, who did have the requisite license.

3. A stone inscription on the wall of the MNI dated 1953 refers to this grand mapping project: "Where shall wisdom be found and where is the place of understanding." This is presumably based on Job 28:12, "But where shall wisdom be found? and where is the place of understanding?" in the King James version.

4. This is Artifact 2002.0075.001 listed by the Canada Science and Technology Museums Corporation, listed online at https://ingeniumcanada.org.

5. Mo Costandi's "Wilder Penfield, Neural Cartographer" (2008) blog entry offers an excellent journalistic description of the neurosurgical procedure, along with further detailed explanation of the neuroscience.

6. Letter from Wilder Penfield to Francis Walshe, August 20, 1946. Source: Penfield archive, Osler Library, McGill University.

2. Pain as a Distinct Sensation

1. In fact, Hardcastle and other print and online sources that replicate this wording have misquoted Faulkner. The final line of the novel is: "Between grief and nothing I will take grief" (Faulkner [1939] 1995, 273). For the purposes of this chapter, the misquotation works better.

2. The prefix "noci-" derives from the Latin *nocēre*, "to hurt, injure" (OED online 2017). Thus the related term that Sherrington uses, "noxious" (meaning harmful or injurious), is derived from the same root. Source: "noxa, n." *OED Online*, Oxford University Press, June 2017, www.oed.com/view/Entry/128855.

4. "The Neuro-motor Unconscious"

1. Martha Nussbaum's very comprehensive translation and interpretation of Aristotle's text *Peri zoon kinēseos* (On the movement of animals) explains that the authorship was disputed for a while, and not always included in Arabic compendia

of Aristotelian texts. Hence its philosophical importance was downplayed. However, Nussbaum agrees with other scholars that it was likely written during Aristotle's second visit to Athens, c335–323 BCE (Nussbaum 1985, 12).

2. Since no English translation of Marey's *La méthode graphique dans les sciences expérimentales et principalement en physiologie et en médecine* (The graphic method in the experimental sciences and more specifically in physiology and medicine) exists, wherever quoted I must rely on fragments translated by others.

3. Anne Hollander's review of Solnit's initial hardback edition of her book on Muybridge, with a different title, was published in the *London Review of Books* (Hollander 2003) in July. A letter published in the very next issue by Brian Winston takes Solnit to task for some perceived inaccuracies, including Muybridge's role as the "father" of motion pictures. However, along the way Winston introduces the fact that du Hauron conceived of and patented the first motion picture apparatus in 1864, "which used up to 580 lenses to capture motion" (Winston 2003). I can find no other corroboration of this number of lenses (see, e.g., Forch 1920) and must consider this a mistake.

4. Dierig's (2003) larger argument about the rise of the "networked city organism" in the industrial nineteenth century, and the rise of these urban physiological laboratories, places machine power and the organization of labor power as a common center. Part of his narrative of the "laboratory revolution" is that laboratory work itself becomes industrialized: "Large-scale, cooperative knowledge production of the physiology institute replace[s] small-scale, solitary production of the modest laboratory," he argues (133).

5. There is some dispute in the secondary literature over the actual title of Marey's Chair at the Collège de France and the date of his accession. Rabinbach (1992, 89) has Marey's appointment in 1866 as "natural history of organized bodies," whereas Braun (1994, 39) has Marey's appointment in 1869 as "history of organized bodies." In an appendix listing appointments at the Collège de France, Appel (1987, 241) lists the Chair as "natural history of organized bodies," and the year 1869.

6. Dagognet (1992, 29) has *kūma* as "swell," "wave," and *The American Illustrated Medical Dictionary* of 1916 also has the derivation "wave" under the entry "kymograph" (Dorland 1916, 508).

7. Originally a lecture delivered at the Collège de France, Marey's text "La station physiologique de Paris" was printed in two parts in *Revue Scientifique*: in volume 2 (December 29, 1894, 802–8), and volume 3 (January 8, 1895, 2–12). The text was anonymously translated into English for the Smithsonian Institution report of activities to July 1894, printed in 1896.

BIBLIOGRAPHY

Abelson, Abraham R. (1908) 1911. "Mental Fatigue and Its Measurement by the Aesthe-siometer." In *Internationales Archiv für Schulhygiene,* vol. 5, edited by L. Brunton, 347–87. Leipzig, Ger.: Verlag von Wilhelm Engelmann.

Alba, Davey. 2016. "Obama Geeks Out Over a Brain-Controlled Robotic Arm That 'Feels.'" Wired. October 14, 2016. https://www.wired.com/2016/10/obama-geeks-brain-controlled-robotic-arm-feels/.

Allen, Grant. 1877. *Physiological Aesthetics.* New York: D. Appleton.

Allison, T., G. McCarthy, M. Luby, A. Puce, and D. D. Spencer. 1996. "Localization of Functional Regions of Human Mesial Cortex by Somatosensory Evoked Potential Recording and by Cortical Stimulation." *Electroencephalography and Clinical Neurophysiology* 100:126–40.

Amar, Jules. 1920. *The Human Motor; or, The Scientific Foundations of Labour and Industry.* Translated by E. P. Butterworth and G. E. Wright. London: George Routledge and Sons.

Anonymous. 1922. "The British Industrial Fatigue Research Board." *Science,* n.s., 55 (1423): 368–69.

Anonymous. 1893. Philosophical Periodicals. *Mind,* n.s., 2 (8): 552–56.

Appel, Toby A. 1987. *The Cuvier-Geoffrey Debate: French Biology in the Decades before Darwin.* New York: Oxford University Press.

Aristotle. 1984. *Sense and Sensibilia [De Sensu].* Translated by J. I. Beare. In *Aristotle: Works,* vol. 2, edited by W. D. Ross, 693–713. Princeton, N.J.: Princeton University Press.

Auvray, Malika, Erik Myin, and Charles Spence. 2010. "The Sensory-Discriminative and Affective-Motivational Aspects of Pain." *Neuroscience & Biobehavioral Reviews* 34 (2): 214–23.

Baldwin, James M, ed. 1905. *Dictionary of Philosophy and Psychology.* New York: MacMillan.

Ball, Edward. 2013. *The Inventor and the Tycoon: The Murderer Eadweard Muybridge, the Entrepreneur Leland Stanford, and the Birth of Moving Pictures.* New York: Anchor Books.

Bastian, Henry C. 1880. *The Brain as an Organ of Mind.* London: C. Kegan Paul.

Bastian, Henry C. 1887. "The 'Muscular Sense'; Its Nature and Cortical Localisation." *Brain* 10:1–137.

Bates, William N. 1919. "Archaeological Discussions: Summaries of Original Articles Chiefly in Current Publications." *American Journal of Archaeology* 24 (2): 173–274.

Bateson, Gregory. 1972. *Steps to an Ecology of Mind.* New York: Ballantine Books.

Batten, Fred E. 1897. "The Muscle-Spindle under Pathological Conditions." *Brain* 20 (1–2): 138–79.

Bauman, Zygmunt. 2000. *Liquid Modernity*. Cambridge, UK: Polity Press.

Bell, Charles. 1823. "On the Motions of the Eye, in Illustration of the Uses of the Muscles and the Orbit." *Philosophical Transactions of the Royal Society of London* 113:166–86.

Bell, Charles. 1826. "On the Nervous Circle Which Connects the Voluntary Muscles with the Brain." *Philosophical Transactions of the Royal Society of London* 116:163–73.

Bell, Charles. 1833. *The Hand: Its Mechanism and Vital Endowments as Evincing Design*. London: W. Pickering.

Bendelow, Gillian, and Simon Williams. 1995. "Pain and the Mind-Body Dualism: A Sociological Approach." *Body and Society* 1 (2): 83–103.

Benjamin, Walter. 2006. *The Writer of Modern Life: Essays on Charles Baudelaire*. Edited by Michael W. Jennings. Cambridge, Mass.: Harvard University Press.

Bennett, Max R., and Peter M. S. Hacker. 2008. *History of Cognitive Neuroscience*. Malden, Mass.: Wiley-Blackwell.

Berenson, Bernard. 1906. *The Florentine Painters of the Renaissance*. 3rd ed. London: G. P. Putnam and Sons.

Berenson, Bernard. 1948. *Aesthetics and History in the Visual Arts*. New York: Pantheon.

Bergson, Henri. (1907) 1911. *Creative Evolution*. Translated by Arthur Mitchell. New York: Henry Holt.

Bergson, Henri. (1896) 1999. *Matter and Memory*. Translated by N. M. Paul and W. S. Palmer. New York: Zone Books.

Bernard, Claude. (1865) 1984. *Introduction a l'étude de la médecine expérimentale*. Paris: Flammarion.

Bernhardt, Martin. 1872. "Zur Lehrer von Muskelsinn." *Archiv für Psychiatrie und Nervenkrankheiten* 3 (3): 618–35.

Berthoz, Alain. 1991. "Reference Frames for the Perception and Control of Movement." In *Brain and Space*, edited by J. Paillard, 81–111. Oxford: Oxford University Press.

Bichat, Marie Francois Xavier. (1800) 1815. *Physiological Researches on Life and Death*. Translated by F. Gold. London: Longman, Hurst, Rees, Orme and Browne.

Binet, Alfred, and Jules Courtier. 1897. "Effets du travail intellectuel sur la circulation capillaire." *L'Année Psychologique* 3:42-64.

Binet, Alfred, and Nicolae Vaschide. 1897. "The Influence of Intellectual Work on the Blood Pressure of Man." *Psychological Review* 4 (1): 54–66.

Binet, Alfred, and Victor Henri. 1898. *La Fatigue intellectuelle*. Paris: Schleicher.

Blair, A. 1999. "Authorship in the Popular 'Problemata Aristotelis.'" *Early Science and Medicine* 4 (3): 189–227.

Boldrey, Edwin B. 1936. "The Architectonic Subdivision of the Mammalian Cerebral Cortex: Including a Report of Electrical Stimulation of One Hundred and Five Human Cerebral Cortices." Master's thesis, McGill University.

BMJ. 1884. "The Functions of the Cerebellum." *British Medical Journal* 2 (1249): 1147–48.

BMJ. 1900. "Orientation." *British Medical Journal* 2 (2080): 1393–94.

BMJ. 1921. "Industrial Fatigue Research Board." *British Medical Journal* 1 (3136): 203–5.

BMJ. 1922. "Industrial Fatigue Research Board." *British Medical Journal* 1 (3191): 322.

BMJ. 1924. "Industrial Fatigue Research Board." *British Medical Journal* 2 (4263): 349.

Borch, Christian. 2008. *Architectural Atmospheres*. Basel, Switz.: Birkhäuser Verlag.

Borell, Merriley. 1993. "Training the Senses, Training the Mind." In *Medicine and the Five Senses*, edited by W. F. Barnum and R. Porter, 244–61. Cambridge: Cambridge University Press.

Boring, Edwin G. 1942. *Sensation and Perception in the History of Experimental Psychology*. Oxford: D. Appleton-Century.

Brain, Robert M. 2008. "The Pulse of Modernism: Experimental Physiology and Aesthetic Avant-Gardes Circa 1900." *Studies in History and Philosophy of Science* 39:393–417.

Brain, Robert M. 2016. *The Pulse of Modernism: Physiological Aesthetics in Fin-de-Siècle Europe.* Seattle: University of Washington Press.

Braun, Marta. 1994. *Picturing Time: The Work of Étienne-Jules Marey (1830–1904).* Chicago: University of Chicago Press.

Braun, Marta, and Elizabeth Whitcombe. 1999. "Marey, Muybridge, and Londe." *History of Photography* 23 (3): 218–24.

Broca, Paul. 1861. "Sur le principe des localisations cérébrales." *Bulletin de la Société d'Anthropologie* 2:190–204.

Brookman, Philip. 2010. *Helios: Eadweard Muybridge in a Time of Change.* New York: Steidl.

Brown, Thomas. 1820. *Lectures on the Philosophy of the Human Mind.* 4 Volumes. Edinburgh: William Tait.

Burke, Robert E. 2007. "Sir Charles Sherrington's 'The Integrative Action of the Nervous System': A Centenary Appreciation." *Brain* 130:887–94.

Burr, David C., and Concetta Morrone. 2004. "Visual Perception during Saccades." In *The Visual Neurosciences,* edited by J. S. Werner and L. M. Chalupa, 1391–401. London: MIT Press.

Cache, Bernard. 1995. *Earth Moves: The Furnishing of Territories.* Edited by M. Speaks. Translated by A. Boyman. London: MIT Press.

Canales, Jimena. 2009. *A Tenth of a Second: A History.* Chicago: University of Chicago Press.

Candlin, Fiona. 2006. "The Dubious Inheritance of Touch: Art History and Museum Access." *Journal of Visual Culture* 5 (2): 137–54.

Candlin, Fiona. 2010. *Art, Museums and Touch.* Manchester: Manchester University Press.

Canguilhem, Georges. (1947) 1992. "Machine and Organism." Translated by Mark Cohen and Randall Cherry. In *Incorporations,* edited by Jonathan Crary and Sanford Kwinter, 45–65. New York: Zone Books.

Cannon, Walter Bradford. 1939. *The Wisdom of the Body.* New York: W. W. Norton.

Carpenter, William Benjamin. 1888. *Nature and Man: Essays Scientific and Philosophical.* London: Kegan and Paul.

Carter, Ian. 1997. "Rain, Steam and What?" *Oxford Art Journal* 20 (2): 3–12.

Catani, Marco. 2017. "A Little Man of Some Importance." *Brain* 140 (11): 3055–61.

Clark, Liat. 2014. "Men Would Rather Receive an Electric Shock Than Think." *Wired.* July 4, 2014. http://www.wired.co.uk/article/electric-shock-therapy-better-than -thinking.

Classen, Constance. 1993. *Worlds of Sense: Exploring the Senses in History and across Cultures.* London: Routledge.

Classen, Constance, ed. 2005. *The Book of Touch.* Oxford: Berg.

Classen, Constance. 2012. *The Deepest Sense: A Cultural History of Touch.* Chicago: University of Illinois Press.

Coats, Joseph. 1875. "Exchange Journals: Reichert and Du Bois-Reymond's *Archiv.*" *Glasgow Medical Journal,* n.s., 7:378–94.

Cole, Jonathan. 1995. *Pride and a Daily Marathon.* Cambridge, Mass.: MIT Press.

Costandi, Mo. 2008. "Wilder Penfield, Neural Cartographer." *Neurophilosophy* (blog). August 27, 2008. https://neurophilosophy.wordpress.com/2008/08/27/wilder _penfield_neural_cartographer/.

Craig, Arthur D. 2002. "How Do You Feel? Interoception: The Sense of the Physiological Condition of the Body." *Nature Reviews Neuroscience* 3:655–66.

Craig, Arthur D. 2003. "A New View of Pain as a Homeostatic Emotion." *Trends in Neurosciences* 26 (6): 303–7.

Crary, Jonathan. 1992. *Techniques of the Observer: On Vision and Modernity in the Nineteenth Century.* Cambridge, Mass.: MIT Press.

Crary, Jonathan. 2001. *Suspensions of Perception: Attention, Spectacle, and Modern Culture.* Cambridge, Mass.: MIT Press.

Cullen, Kathleen E. 2012. "The Vestibular System: Multimodal Integration and Encoding of Self-Motion for Motor Control." *Trends in Neurosciences* 35 (3): 185–96.

D.S. 1944. "Absence from Work: Prevention of Fatigue, Industrial Health Research Board Pamphlet No. 2." Review essay. *British Journal of Industrial Medicine* 1 (3): 201.

Dagognet, François. 1992. *Étienne-Jules Marey: A Passion for the Trace.* New York: Zone.

Dallenbach, Karl M. 1939. "Pain: History and Present Status." *American Journal of Psychology* 52 (3): 331–47.

Damasio, Antonio. 1994. *Descartes' Error: Emotion, Reason, and the Human Brain.* New York: G. P. Putnam's Sons.

Danius, Sarah. 2002. *The Senses of Modernism: Technology, Perception and Aesthetics.* Ithaca, N.Y.: Cornell University Press.

DARPA. N.d. "DARPA and the Brain Initiative." Accessed April 22, 2020. https://www.darpa.mil/program/our-research/darpa-and-the-brain-initiative.

Daston, Lorraine, and Peter Galison. 1992. "The Image of Objectivity." *Representations* 40:81–128.

de Chadarevian, Soraya. 1993. "Graphical Method and Discipline: Self-Recording Instruments in Nineteenth-Century Physiology." *Studies in History and Philosophy of Science Part A* 24 (2): 267–91.

Deleuze, Gilles, and Félix Gauttari. 1988. *A Thousand Plateaus: Capitalism and Schizophrenia.* Translated by B. Massumi. London: Athlone.

DeSalle, Robert. 2018. *Our Senses: An Immersive Experience.* New Haven, Conn.: Yale University Press.

Dewey, James. 1896. "The Reflex Arc Concept in Psychology." *Psychological Review* 3:357–70.

Dierig, Sven. 2003. "Engines for Experiment: Laboratory Revolution and Industrial Labor in the Nineteenth-Century City." *Osiris,* 2nd ser., 18:116–34.

Di Giulio, Camillo, Franca Daniele, and Charles M. Tipton. 2006. "Angelo Mosso and Muscular Fatigue: 116 Years after the First Congress of Physiologists; IUPS Commemoration." *Advances in Physiology Education* 30:51–57.

Dijkerman, H. Chris, and Edward H. F. de Haan. 2007. "Somatosensory Processes Subserving Perception and Action." *Behavioral and Brain Sciences* 30:189–239.

Di Noto, Paula M., Leorra Newman, Shelley Wall, and Gillian Einstein. 2013. "The *Her*munculus: What Is Known about the Representation of the Female Body in the Brain?" *Cerebral Cortex* 23:1005–13.

Doane, Mary Ann. 2002. *The Emergence of Cinematic Time: Modernity, Contingency, The Archive.* Cambridge, Mass.: Harvard University Press.

Dorland, W. A. Newman. 1916. "Kymograph." In *The American Illustrated Medical Dictionary,* 508. 8th ed. Philadelphia: W. B. Saunders.

Drucker, Johanna. 1996. *Theorizing Modernism: Visual Art and the Critical Tradition.* New York: Columbia University Press.

Edwards, Richard H. T., Veronica Toescu, Henry Gibson. 2013. "Historical Perspective: A Framework for Interpreting Pathobiological Ideas on Human Muscle Fatigue." In *Fatigue: Neural and Muscular Mechanisms (Advances in Experimental Medicine and Biology)*, edited by Roger M. Enoka et al., 481–94. New York: Springer.

Eisler, Rudolph. 1904. *Wörterbuch der philosophischen Begriffe*. Vol. 1. Berlin: E. S. Mittler und Sohn.

Elkins, James. 1994. *The Poetics of Perspective*. Ithaca, N.Y.: Cornell University Press.

Erickson, T. C. 1945. "Erotomania as an Expression of Cortical Epileptiform Discharge." *Archives of Neurological Psychiatry* 53:226–31.

Fagan, Garrett G. 1999. *Bathing in Public in the Roman World*. Ann Arbor: University of Michigan Press.

Faulkner, William. (1939) 1995. *The Wild Palms*. New York: Vintage International.

Fechner, Gustav Theodor. (1860) 1996. *Elemente der Psychophysik*. Leipzig, Ger.: Von Breitkopf und Härtel.

Feindel, William. 1962. "Thomas Willis (1621–1675): The Founder of Neurology." *Canadian Medical Association Journal* 87 (6): 289–96.

Feindel, William, and Richard Leblanc. 2016. *The Wounded Brain Healed: The Golden Age of the Montreal Neurological Institute, 1934–1993*. Montreal: McGill-Queen's Press.

Ferrier, David. 1876. *The Functions of the Brain*. New York: G. P. Putnam's Sons.

Ferrier, David. 1878. "The Goulstonian Lectures on the Localisation of Cerebral Disease." *British Medical Journal* 1 (900): 443–47.

Finger, Stanley. 2001. *Origins of Neuroscience: A History of Explorations into Brain Function*. Oxford: Oxford University Press.

Finger, Stanley. 2005. *Minds behind the Brain: A History of the Pioneers and Their Discoveries*. New York: Oxford University Press.

Flesher, Sharlene N., Jennifer L. Collinger, Stephen T. Foldes, Jeffrey M. Weiss, John Downey, Elizabeth C. Tyler-Kabara, Sliman J. Bensmaia, Andrew B. Schwartz, Michael L. Boninger, and Robert A. Gaunt. 2016. "Intracortical Microstimulation of Human Somatosensory Cortex." *Science Translational Medicine* 8 (361): 361ra141.

Florence, Philip S. (1957) 2003. *Industry and the State*. New York: Routledge.

Foerster, Otfrid. 1936. "The Motor Cortex in Man in Light of Hughlings Jackson's Doctrines." *Brain* 59:136–59.

Forch, Carl. 1920. "Louis Arthur Ducos du Hauron: In Memoriam." *Photo-Era: The American Journal of Photography* 45 (6): 281–82.

Foster, Michael, and Charles S. Sherrington. 1897. *Textbook of Physiology*. Vol. 3. 7th ed. London: Macmillan.

Foster, Susan L. 2011. *Choreographing Empathy: Kinesthesia in Performance*. New York: Routledge.

Foucault, Michel. 2002. *The Archaeology of Knowledge*. Translated by A. M. Sheridan Smith. London: Routledge.

Foucault, Michel. 2003. *The Birth of the Clinic: An Archaeology of Medical Perception*. Translated by A. M. Sheridan. London: Routledge.

Fox, Kieran C. R., Evan Thompson, Jessica Andrews-Hanna, and Kalina Christoff. 2014. "Is Thinking Really Aversive? Commentary on Wilson et al.'s 'Just Think: The Challenges of the Disengaged Mind.'" *Frontiers in Psychology* 5 (1427). http://www.frontiersin.org/Journal/FullText.aspx?s=194&name=cognition&ART_DOI=10.3389/fpsyg.2014.01427.

Fritsch, Gustav T., and Edvard Hitzig. 1870. "On the Electrical Excitability of the

Cerebrum." Translated in G. Von Bonin. 1960. *Some Papers on the Cerebral Cortex,* 73–96. Springfield, Ill.: Charles C. Thomas.

Galison, Peter. 1994. "The Ontology of the Enemy: Norbert Wiener and the Cybernetic Vision." *Critical Inquiry* 21 (1): 228–66.

Gallagher, Shaun. 1986. "Body Image and Body Schema: A Conceptual Clarification." *Journal of Mind and Behavior* 7 (4): 541–54.

Gandhoke, Gurpreet S., Evgenii Belykh, Xiaochun Zhao, Richard Leblanc, and Mark C. Preul. 2019. "Edwin Boldrey and Wilder Penfield's Homunculus: A Life Given by Mrs. Cantlie (In and Out of Realism)." *World Neurosurgery* 132:377–88.

Gatchel, Robert J., Yuan B. Peng, Mandelon L. Peters, Perry N. Fuchs, and Dennis C. Turk. 2007. "The Biopsychosocial Approach to Chronic Pain: Scientific Advances and Future Directions." *Psychological Bulletin* 133 (4): 581–624.

Gebhart, Gerald F., and Robert F. Schmidt. 2013. "Retinal Slip." In *Encyclopedia of Pain,* edited by G. F. Gebhart and R. F. Schmidt, 3407. 2nd ed. Berlin: Springer-Verlag.

Gelb, Adhémar, and Kurt Goldstein. 1920. "Zur Psychologie des optischen Wahr-nehmungs und Erkennungsvorgangs." In *Psychologische Analysen hirnpatholo-gischer Fälle auf Grund von Untersuchungen Hirnverletzter,* edited by A. Gelb and K. Goldstein, 157–250. Leipzig: J. A. Barth, 1920.

George, L. 1870. "Der Muskelsinn." In *Archiv für Anatomie, Physiologie und wissen-schaftliche Medicin,* 251–63. Leipzig, Ger.: Von Veit.

Geschwind, Norman. 1979. "Specializations of the Human Brain." *Scientific American* 241 (3): 180–201.

Geurts, Kathryn L. 2002. *Culture and the Senses: Bodily Ways of Knowing in an African Community.* Berkeley: University of California Press.

Gibson, James J. 1966. *The Senses Considered as Perceptual Systems.* London: George Allan & Unwin.

Gibson, James J. 1979. *The Ecological Approach to Visual Perception.* London: Houghton Mifflin.

Gilbreth, Frank B. 1906. *Bricklaying System.* New York: Myron C. Clark.

Gilbreth, Frank B. 1911. *Motion Study: A Method for Increasing the Efficiency of the Workman.* New York: D. Van Norstrand.

Gilbreth, Frank B., and Lillian M. Gilbreth. 1916. *Fatigue Study: The Elimination of Humanity's Greatest Unnecessary Waste; A First Step in Motion Study.* New York: Sturgis and Walton.

Gilbreth, Frank B., and Lillian M. Gilbreth. 1917. *Applied Motion Study: A Collection of Papers on the Efficient Method to Industrial Preparedness.* New York: Sturgis & Walton.

Gilbreth, Lillian M. 1927. *The Home-Maker and Her Job.* New York: D. Appleton.

Gillespie, Richard. 1987. "Industrial Fatigue and the Discipline of Physiology." In *Physiology in the American Context 1850–1940,* edited by G. L. Geison, 237–62. Baltimore: American Physiological Society.

Goldmark, Josephine C. 1912. *Fatigue and Efficiency: A Study in Industry.* New York: Charities Publication Committee.

Goldscheider, Alfred. 1889. "Über den Muskelsinn und die Theorie der Ataxie." *Zeit-schrift für klinische Medicin* 15:82–161.

Goldstein, Kurt. (1934) 2000. *The Organism: A Holistic Approach to Biology Derived from Pathological Data in Man.* New York: Zone Books.

Gradenwitz, Alfred. 1905. "Improved Methods in High-Speed Chronophotography," *Scientific American* 93 (3): 50–51.

Grahek, Nikola. 2014. *Feeling Pain and Being in Pain.* 2nd ed. Cambridge, Mass.: MIT Press.

Granit, Ragnar. 1973. "Muscle Sense, Proprioception, and the Control of Movement." In *Motor Control,* edited by A. A. Gydikov, N. T. Tankov, and D. S. Kosarov, 1–14. New York: Plenum Press.

Greicius, Michael D., and Vinod Menon. 2004. "Default-Mode Activity during a Passive Sensory Task: Uncoupled from Deactivation but Impacting Activation." *Journal of Cognitive Neuroscience* 16 (9): 1484–92.

Griffin, William. 1826. *An Essay on the Nature of Pain; with Some Considerations on Its Principal Varieties and Connected with Disease, and Remarks on the Treatment.* Edinburgh: J. Moir.

Griggs, Richard A. 1988. "Who Is Mrs. Cantlie and Why Are They Doing Those Terrible Things to Her Homunculi?" *Teaching of Psychology* 15 (2): 105–6.

Gross, Charles G. 2007. "The Discovery of Motor Cortex and Its Background." *Journal of the History of the Neurosciences* 16 (3): 320–31.

Grünbaum, Abraham A. 1930. "Aphasie und Motorik." *Zeitschrift fuer die Gesamte Neurologie und Psychiatrie* 130:385–412.

Guenther, Katja. 2015. *Localization and Its Discontents: A Genealogy of Psychoanalysis and the Neuro Disciplines.* Chicago: University of Chicago Press.

Guenther, Katja. 2016. "Between Clinic and Experiment: Wilder Penfield's Stimulation Reports and the Search for Mind, 1929–55." *Canadian Bulletin of Medical History* 33 (2): 281–320.

Gusnard, Debra A., Erbil Akbudak, Gordon L. Shulman, and Marcus E. Raichle. 2001. "Medial Prefrontal Cortex and Self-Referential Mental Activity: Relation to a Default Mode of Brain Function." *Proceedings of the National Academy of Sciences* 98 (7): 4259–64.

Gusnard, Debra A., and Marcus E. Raichle. 2001. "Searching for a Baseline: Functional Imaging and the Resting Human Brain." *Nature Reviews: Neuroscience* 2:685–94.

Gutman, Sharon A. 2007. *Quick Reference Neuroscience for Rehabilitation Professionals: The Essential Neurologic Principles Underlying Rehabilitation Practice.* 2nd ed. Thorofare, N.J.: Slack Books.

Haas, Robert B. 1972. "Eadweard Muybridge, 1830–1904." In *Eadweard Muybridge: The Stanford Years, 1872–1882,* edited by A. V. Mozley, R. B. Haas, and F. Forster-Hahn, 11–36. Stanford, Calif.: Department of Art, Stanford University.

Hale, Jonathan. 2013. "Critical Phenomenology: Architecture and Embodiment." *Architecture & Ideas* 12:18–37.

Hall, Marshall. 1833. "On the Reflex Functions of the Medulla Oblongata and the Medulla Spinalis." *Philosophical Transactions of the Royal Society B* 123:635–66.

Hamlyn, David W. 1959. "Aristotle's Account of Aesthesis in the *De Anima.*" *Classical Quarterly,* n.s., 9 (1): 6–16.

Hardcastle, Valerie G. 1999. *The Myth of Pain.* London: MIT Press.

Harlow, John Martyn. 1848. "Passage of an Iron Rod through the Head." *Boston Medical & Surgical Journal* 39 (20): 389–93.

Harlow, John Martyn. 1868. "Recovery from the Passage of an Iron Bar through the Head." *Publications of the Massachusetts Medical Society* 2 (3): 327–47.

Harris, Anna. 2016. "Listening-Touch, Affect and the Crafting of Medical Bodies through Percussion." *Body and Society* 22 (1): 31–61.

Hartoonian, Gevork. 2001. "The Limelight of the House-Machine." *Journal of Architecture* 6 (1): 53–79.

Hawkins, Joseph E. 2004. "Sketches of Otohistory Part 1: Otoprehistory; How It All Began." *Audiology & Neurotology* 9 (2): 66–71.

Head, Henry. 1920. *Studies in Neurology.* Oxford: Oxford University Press.

Head, Henry, and Gordon Holmes. 1911. "Sensory Disturbances from Cerebral Lesions." *Brain* 34:102–254.

Heller-Roazen, Daniel. 2007. *The Inner Touch: Archaeology of Sensation.* New York: Zone Books.

Helmholtz, Herman L. von. 1873. "On the Interaction of Natural Forces." In *Popular Lectures on Scientific Subjects,* 153–96. Translated by E. Atkinson. New York: D. Appleton.

Henn, Volker E. 1984. "Mach on the Analysis of Motion Sensation." *Human Neurobiology* 3:145–48.

Henry, Charles. 1885. "Introduction à une ésthetique scientifique." *La revue contemporaine* 2:441–69.

Herder, Johann G. (1778) 2002. *Sculpture: Some Observations on Shape and Form from Pygmalion's Creative Dream.* Translated by J. Gaiger. Chicago: University of Chicago Press.

Hirst, Paul Q. 2010. *Durkheim, Bernard and Epistemology.* London: Routledge.

Hocheisen, Paul. 1893. "Über den Muskelsinn bei Blinden," *Zeitschrift für Psychologie und Physiologie der Sinnesorgane* 5:239–82.

Holl, Stephen, Juhanni Pallasmaa, and Alberto Perez-Gomez, eds. 2006. *Questions of Perception: Phenomenology of Architecture.* San Francisco: William Stout.

Hollander, Anne. 2003. "Bought a Gun, Found the Man." *London Review of Books* 25 (14): 9–12.

Honkasalo, Marja-Liisa. 1998. "Space and the Embodied Experience: Rethinking the Body in Pain." *Body and Society* 4 (2): 35–57.

Horn, David G. 1995. "This Norm Which Is Not One: Reading the Female Body in Lambroso's Anthropology." In *Deviant Bodies: Critical Perspectives on Difference in Science and Popular Culture,* edited by Jennifer Terry and Jacqueline L Urla, 109–28. Indianapolis: Indiana University Press.

Horvath, Steven M., and Elizabeth C. Horvath. 1973. *The Harvard Fatigue Laboratory: Its History and Contributions.* Englewood Cliffs, N.J.: Prentice-Hall.

Howes, David. 2009. "Introduction: The Revolving Sensorium." In *The Sixth Sense Reader,* edited by D. Howes, 1–52. Oxford: Berg.

IFRB. 1920. *First Annual Report of the Industrial Fatigue Research Board to 31st March, 1920.* London: His Majesty's Stationery Office.

IHRB. 1944. *Absence from Work: Prevention of Fatigue.* Industrial Health Research Board Pamphlet No. 2. London: H.M.S.O.

Ingold, Tim. 2004. "Culture on the Ground: The World Perceived through the Feet." *Journal of Material Culture* 9 (3): 315–40.

Ionescu, Vlad. 2016. "Architectural Symbolism: Body and Space in Heinrich Wölfflin and Wilhelm Worringer." *Architectural Histories* 4 (1): 1–9.

JAMA. 1942. "Fatigue of Workers: Its Relation to Industrial Production" (review). *Journal of the American Medical Association* 119 (8): 683.

James, William. 1880. *The Feeling of Effort.* Boston: Boston Society of Natural History.

James, William. 1881. "Notes on a Sense of Dizziness in Deaf-Mutes." *Harvard University Bulletin* 2 (18): 173.

James, William. 1882. "The Sense of Dizziness in Deaf-Mutes." *American Journal of Otology* 4:239–54.

James, William. 1890. *The Principles of Psychology.* 2 vols. New York: Henry Holt.

Jashemski, Wilhelmina F. 1995. "Roman Gardens in Tunisia: Preliminary Excavations in the House of Bacchus and Ariadne and in the East Temple at Thuburbo Maius." *American Journal of Archaeology* 99 (4): 559–76.

Jay, Martin. 1994. *Downcast Eyes: The Denigration of Vision in Twentieth-Century French Thought.* London: University of California Press.

Jeannerod, Marc. 1985. *The Brain Machine: The Development of Neurophysiological Thought.* Cambridge, Mass.: Harvard University Press.

Jensen, Rasmus Thybo. 2009. "Motor Intentionality and the Case of Schneider." *Phenomenology and Cognitive Science* 8:371–88.

Jones, Edward G. 1972. "The Development of the 'Muscular Sense' Concept during the Nineteenth Century and the Work of H. Charlton Bastian." *Journal of the History of Medicine and Allied Sciences* 27 (3): 298–311.

Jütte, Robert. 2005. *A History of the Senses: From Antiquity to Cyberspace.* Translated by J. Lynn. Cambridge, UK: Polity.

Kelly, Sean Dorrance. 2002. "Merleau-Ponty on the Body." *Ratio,* n.s., 15:376–91.

Kemp, Simon, and Garth J. O. Fletcher. 1993. "The Medieval Theory of the Inner Senses." *American Journal of Psychology* 106 (4): 559–76.

Kirchner, Freidrich. 1886. *Wörterbuch der philosophischen Grundbegriffe.* Heidelberg, Ger.: Weiss.

Kirchner, Freidrich. 1888. *A Student's Manual of Psychology: Adapted from the "Katechismus Der Psychologie" of Friedrich Kirchner.* Translated and edited by E. D. Drought. London: Swan Sonnenschein, Lowrey.

Kirchner, Freidrich. 1907. *Wörterbuch der philosophischen Grundbegriffe.* Rev. ed. Edited by Dr. Carl Michaëlis. Leipzig, Ger.: Verlag de Dürr'schen Buchhandlung.

Klein, E., J. A. Langley, and E. A. Schäfer. 1883. "Report on the Cortical Areas Removed from the Brain of a Dog, and from the Brain of a Monkey." *Journal of Physiology* 4 (4–5): 231–47.

Komisaruk, Barry R., Nan Wise, Eleni Frangos, Wen-Ching Liu, Kachina Allen, and Stuart Brody. 2011. "Women's Clitoris, Vagina, and Cervix Mapped on the Sensory Cortex: fMRI Evidence." *Journal of Sexual Medicine* 8 (10): 2822–30.

Kraepelin, Emil. 1902. "Die Arbeitscurve." *Philosophische Studien* 19:459–507.

Krause Fedor. 1908–11. *Chirurgie des Gehirns und Rückenmarks nach eigenen Erfahrungen.* Vols. 1 and 2. Berlin: Urban und Schwarzenburg.

Krause Fedor. 1912. *Surgery of the Brain and Spinal Cord Based on Personal Experience.* Vol. 2. Translated by Max Thorek. New York: Rebman.

Kronecker, Hugo. 1871. "Über die Ermüdung und Erholung der quergestreiften Muskeln." *Arbeiten aus der Physiologischen Anstalt du Leipzig* 6:177–266.

Lackner, James R., and Paul DiZio. 2005. "Vestibular, Proprioceptive, and Haptic Contributions to Spatial Orientation." *Annual Review of Psychology* 56:115–47.

Lalvani, Suren. 1996. *Photography, Vision, and the Production of Modern Bodies.* Albany: State University of New York Press.

Landes, Donald. 2012. "Translator's Introduction." In Maurice Merleau-Ponty, *Phenomenology of Perception,* xxx–li. New York: Routledge.

Landes, Donald. 2013. *The Merleau-Ponty Dictionary.* New York: Bloomsbury.

Latsis, Dominic. 2015. "Landscape in Motion: Muybridge and the Origins of Chronophotography." *Film History* 27 (3): 1–40.

Leahy, Helen R. 2012. *Museum Bodies: The Politics and Practices of Visiting and Viewing.* Farnham, UK: Ashgate.

Le Corbusier. 1986. *Towards a New Architecture.* New York: Dover.

Lewes, George H. 1878. "Motor-Feelings and the Muscular Sense." *Brain* 1:14–28.

Leyton, A. S. F., and Charles S. Sherrington. 1917. "Observations on the Excitable Cortex of the Chimpanzee, Orang-Utan, and Gorilla." *Experimental Physiology* 11:135–222.

Lincoln, D. F., and S. G. Webber. 1871. "Chronicle I: 3—The Muscular Sense *(Muskelsinn)*." *Journal of Psychological Medicine* 5:396–98.

Lombroso-Ferrero, Gina. (1911) 1972. *Criminal Man: According to the Classification of Cesare Lombroso.* London: G. P. Putnam's Sons.

Luciani, Luigi. 1917. *Human Physiology.* Vol. 4, *The Sense Organs,* edited by G. M. Holmes. Translated by F. A. Welby. New York: MacMillan.

Lupton, Deborah. 2016. *The Quantified Self.* Oxford: Polity.

Macarthur, John. 2008. "Movement and Tactility: Wölfflin and Benjamin on Imitation in Architecture." *Journal of Architecture* 12 (5): 477–87.

Mach, Ernst. (1875) 2001. *Fundamentals of Movement Perception.* Translated by Laurence R. Young, Volker Henn, and Hansjörg Scherberger. New York: Kluwer Academic.

Mach, Ernst. 1875. *Grundlinien der Lehre von den Bewegungsempfindungen.* Leipzig, Ger.: Verlag von Wilhelm Engelmann.

Mach, Ernst. (1896) 1914. *The Analysis of Sensations and the Relation of the Physical to the Psychical.* Translated by C. M. Williams. Chicago: The Open Court.

Macmillan, Malcolm. 2000. *An Odd Kind of Fame: Stories of Phineas Gage.* Cambridge, Mass.: MIT Press.

Macmillan, Malcolm. 2008. "Phineas Gage: Unravelling the Myth." *Psychologist* (British Psychological Society) 21 (9): 828–31. http://thepsychologist.bps.org.uk/volume-21/edition-9/phineas-gage-unravelling-myth.

Marey, Étienne-Jules. 1874. *Animal Mechanism: A Treatise on Terrestrial and Aërial Locomotion.* New York: D. Appleton.

Marey, Étienne-Jules. 1878. *La Méthode graphique dans les sciences expérimentales et principalement en physiologie et en médecine.* Paris: Masson.

Marey, Étienne-Jules. 1882. "La fusil photographique." *La Nature* 464 (April 22): 326–30.

Marey, Étienne-Jules. 1883. "La station physiologique de Paris." *La Nature* 536 (September 8): 226–30; and 539 (September 29): 275–79.

Marey, Étienne-Jules. 1894a. "La station physiologique de Paris." *Revue Scientifique* 2 (December 29): 802–8; and 3 (January 8, 1895): 2–12.

Marey, Étienne-Jules. 1894b. *Le mouvement.* Paris: Masson.

Marey, Étienne-Jules. 1895. *Movement.* Translated by Eric Pritchard. New York: D. Appleton.

Marey, Étienne-Jules. 1896. "The Work of the Physiological Station at Paris." In the *Annual Report of the Board of Regents of the Smithsonian Institution,* 391–412. Washington, D.C.: Government Printing Office.

Marey, Étienne-Jules. 1898. "Analyse des mouvements du cheval par la chronophotographie." *La Nature* 1306 (June 11): 22–26.

Marey, Étienne-Jules. 1902. "The History of Chronophotography." In the *Smithsonian Institute, Annual Report 1901,* 317–40. Washington, D.C.: Government Printing Office.

Martin, John J. 1936. *America Dancing: The Background and Personalities of the Modern Dance.* New York: Dodge.

Mauss, Marcel. (1934) 1992. "Techniques of the Body." In *Zone 6: Incorporations,* edited by J. Crary and S. Kwinter, 455–77. New York: Zone.

Mayer, Andreas. 2010. "The Physiological Circus: Knowing, Representing, and Training Horses in Motion in Nineteenth-Century France." *Representations* 111:88–120.

Mazzola, Laure, Jean Isnard, Roland Peyron, and François Mauguière. 2012. "Stimulation of the Human Cortex and the Experience of Pain: Wilder Penfield's Observations Revisited." *Brain* 135 (2): 631–40. https://doi.org/10.1093/brain/awr265.

McComas, Alan J. 2003. "The Neuromuscular System." In *Exercise Physiology: People and Ideas*, edited by Charles M. Tipton, 39–97. New York: Springer.

McIvor, Arthur J. 1987. "Manual Work, Technology, and Industrial Health, 1918–39." *Medical History* 31:160–89.

Melzack, Ronald. 1990. "Phantom Limbs and the Concept of a Neuromatrix." *Trends in Neurosciences* 13 (3): 88–92.

Melzack, Ronald, and W. S. Torgerson. 1971. "On the Language of Pain." *Anesthesiology* 34:50–59.

Melzack, Ronald, and Patrick D. Wall. 1965. "Pain Mechanisms: A New Theory." *Science* 150 (3699): 971–79.

Merleau-Ponty, Maurice. 1964. *The Primacy of Perception*. Translated by J. Edie. Evanston, Ill.: Northwestern University Press.

Merleau-Ponty, Maurice. 1992. *Phenomenology of Perception*. Translated by Colin Smith. London: Routledge.

Merleau-Ponty, Maurice. 2012. *Phenomenology of Perception*. Translated by Donald Landes. New York: Routledge.

Merlin, Alfred. 1917. "Fouilles à Thuburbo Majus en 1916." *Comptes rendus des séances de l'Académie des Inscriptions et Belles-Lettres* 61 (2): 66–77.

Milar, Katherine S. 2012. "William James and the Sixth Sense." *Monitor on Psychology* 43 (8): 22.

Mitchell, Silas Weir. (1872) 1965. *Injuries of Nerves and Their Consequences*. New York: Dover.

Molholt, Rebecca. 2011. "Roman Labyrinth Mosaics and the Experience of Motion." *Art Bulletin* 93 (3): 287–303.

Molnár, Zoltán, and Richard E. Brown. 2010. "Insights into the Life and Work of Sir Charles Sherrington." *Nature Reviews Neuroscience* 11 (6): 429–36.

Montero, Barbera. 2006. "Proprioception as an Aesthetic Sense." *Journal of Aesthetics and Art Criticism* 64 (2): 231–42.

Mooney, Timothy. 2011. "Plasticity, Motor Intentionality and Concrete Movement in Merleau-Ponty." *Continental Philosophy Review* 44: 359–81.

Mososco, Javier. 2012. *Pain: A Cultural History*. Basingstoke, UK: Palgrave.

Mosso, Angelo. (1891) 1904. *Fatigue*. Translated by M. Drummond and W. B. Drummond. New York: G. P. Putnam's Sons.

Mosso, Angelo. 1895. "Sphygmomanomètre pour mesurer la pression du sang chez l'homme." *Archives Italiennes de Biologie* 23 (1): 177–97.

Mozley, Anita V. 1972. "Photographs by Muybridge, 1872–1880: Catalog and Notes on the Work." In *Eadweard Muybridge: The Stanford Years 1872–1882,* edited by A. V. Mozley, R. B. Haas, and F. Forster-Hahn, 37–84. Stanford, Calif.: Department of Art, Stanford University.

Murray, Elsie. 1909. "Organic Sensation." *American Journal of Psychology* 20 (3): 386–446.

Musser, Charles. 2005. "A Cornucopia of Images." In *Moving Pictures: American Art and Early Film, 1880–1910,* edited by Nancy M. Mathews, 5–38. Manchester, Vt.: Hudson Hills Press.

Muybridge, Eadweard. 1955. *The Human Figure in Motion*. Mineola, N.Y.: Dover.

Myers, Charles S. 1925. "The Barbeian Lecture on Industrial Fatigue." *Lancet* 205 (5305): 905–9.

New York Times. 1921. "Working with the Human Motor," July 31, 1921.

Newhall, Beaumont. 1976. *The Daguerreotype in America*. 3rd ed. New York: Dover.

Nicholson, Malcolm. 1993. "The Introduction of Percussion and Stethoscopy to Early Nineteenth-Century Edinburgh." In *Medicine and the Five Senses*, edited by W. F. Barnum and R. Porter, 134–53. Cambridge: Cambridge University Press.

Nicolas, Serge, and Dominique Makowski. 2016. "Can Mental Fatigue Be Measured by Weber's Compass? Alfred Binet's Answer on the Value of Aesthesiometry (Tactile Sensitivity) as an Objective Measure of Mental Fatigue." In *European Yearbook of the History of Psychology*, vol. 2, edited by Mauro Antonelli, 11–46. Turnhout, Bel.: Brepols.

Norrsell, Ulf, Stanley Finger, and Clara Lajonchere. 1999. "Cutaneous Sensory Spots and the 'Law of Specific Nerve Energies': History and Development of Ideas." *Brain Research Bulletin* 48 (5): 457–65.

Norton Wise, Matthew. 2010. "What's in a Line?" In *Cultures and Politics of Research from the Early Modern Period to the Age of Extremes*, edited by Moritz Epple and Claus Zittel, 61–102. Vol. 1 of *Science as Cultural Practice*. Berlin: Akademie Verlag, GmbH.

NRC Committee on Work in Industry. 1941. *Fatigue of Workers: Its Relation to Industrial Production*. New York: Reinhold.

Nussbaum, Martha C. 1985. *Aristotle's De Motu Animalium: Text with Translation, Commentary, and Interpretive Essays*. Princeton, N.J.: Princeton University Press.

Nutt, Amy E. 2016. "In a Medical First, Brain Implant Allows Paralyzed Man to Feel Again." *Washington Post*, October 14, 2016.

Osterhammel, P., K. Terkildsen, and K. Zilstorff. 1968. "Vestibular Habituation in Ballet Dancers." *Acta Oto-laryngologica* 66:221–28.

OED. 2015. "motility, n." OED Online. Accessed August 13, 2015. http://www.oed.com.pitt.idm.oclc.org/view/Entry/122691.

OED. 2015b. "motricity, n." OED Online. Accessed August 13, 2015. http://www.oed.com.pitt.idm.oclc.org/view/Entry/122763.

Pallasmaa, Juhani. 1998. "Logic of the Image." *Journal of Architecture* 3 (1): 289–99.

Pallasmaa, Juhani. 2005. *The Eyes of the Skin: Architecture and the Senses*. London: Academy Editions.

Pallasmaa, Juhani. 2006. "An Architecture for the Seven Senses." In *Questions of Perception: Phenomenology of Architecture*, edited by S. Holl, J. Pallasmaa, and A. P. Gómez, 28–37. San Francisco: William Stout.

Panofsky, Erwin. (1927) 1991. *Perspective as Symbolic Form*. Translated by C. S. Wood. New York: Zone Books.

Parisi, David. 2011. "Tactile Modernity: On the Rationalization of Touch in the Nineteenth Century." In *Media, Technology, and Literature in the Nineteenth Century*, edited by C. Colligan and M. Linley, 189–214. Farnham, UK: Ashgate.

Paterson, Mark. 2007. *The Senses of Touch: Haptics, Affects & Technologies*. Oxford: Berg.

Paterson, Mark. 2009. "Haptic Geographies: Ethnography, Haptic Knowledges and Sensuous Dispositions." *Progress in Human Geography* 33 (6): 766–88.

Paterson, Mark. 2011. "More-Than Visual Approaches to Architecture: Vision, Touch, Technique." *Social & Cultural Geography* 12 (3): 263–81.

Paterson, Mark. 2012. "Movement for Movement's Sake: On the Relationship between Kinesthesia and Aesthetics." *Essays in Philosophy* 13 (2): 471–97.

Paterson, Mark. 2013. "On 'Inner Touch' and the Moving Body: Aisthêsis, Kinaesthesis, and Aesthetics." In *Touching and to be Touched: Kinesthesia and Empathy in Dance*, edited by S. Zubarik and G. Ekert, 115–31. Berlin: Walter de Gruyter.

Paterson, Mark. 2015. "On Aisthêsis, 'Inner Touch,' and the Aesthetics of the Moving Body." In *Geographical Aesthetics: Imagining Space, Staging Encounters*, edited by H. Hawkins and E. Straughan, 35–52. Aldershot, UK: Ashgate.

Paterson, Mark. 2016. *Seeing with the Hands: Blindness, Vision and Touch after Descartes.* Edinburgh: Edinburgh University Press.

Paterson, Mark. 2017. "On Haptics, Tactile Interactions, and the Possibility of a Distinctly 'Haptic Media.'" *New Media and Society* 19 (10): 1541–62.

Paterson, Mark. 2018a. "Motricité, Physiology, and Modernity in *Phenomenology of Perception*." In *Understanding Merleau-Ponty, Understanding Modernism*, 170–184. London: Bloomsbury Academic.

Paterson, Mark. 2018b. "The Biopolitics of Sensation, Techniques of Quantification, and the Production of a 'New' Sensorium." *Resilience: A Journal of the Environmental Humanities* 5 (3): 67–95.

Peña, Carolyn Thomas de la. 2003. *The Body Electric: How Strange Machines Built the Modern American.* New York: New York University Press.

Pencavel, John. 2014. "The Productivity of Working Hours." Forschungsinstitut zur Zukunft der Arbeit [Institute for the Study of Labor]. Discussion Paper no. 8129. http://ftp.iza.org/dp8129.pdf.

Penfield, Wilder. 1930. "The Structural Basis of Traumatic Epilepsy and Results of Radical Operation." *Brain* 53:99–119.

Penfield, Wilder, and Edwin Boldrey. 1937. "Somatic Motor and Sensory Representation in the Cerebral Cortex of Man as Studied by Electrical Stimulation." *Brain* 60 (4): 389–443.

Penfield, Wilder, and H. Jasper. 1954. *Epilepsy and the Functional Anatomy of the Human Brain.* Boston: Little, Brown.

Penfield, Wilder, and Kristian Kristiansen. 1951. *Epileptic Seizure Patterns: A Study of the Localizing Value of Initial Phenomena in Focal Cortical Seizures.* Oxford: Charles C. Thomas.

Penfield, Wilder, and T. Rasmussen. 1950. *The Cerebral Cortex of Man: A Clinical Study of Localization of Function.* New York: Macmillan.

Pfeifer, Rolf, and Josh Bongard. 2007. *How the Body Shapes the Way We Think: A New View of Intelligence.* Cambridge, Mass.: MIT Press.

Phippen, J. Weston. 2016. "The Man Who Captured Time." *The Atlantic,* July 24, 2016. https://www.theatlantic.com/entertainment/archive/2016/07/eadweard -muybridge/483381/.

Pickren, Wade E. 2003. "Kurt Goldstein: Clinician and Philosopher of Human Nature." In *Portraits of Pioneers in Psychology,* vol. 5, edited by Gregory A. Kimble and Michael Wertheimer, 127–40. Mahwah, N.J.: Lawrence Erlbaum.

Pitts-Taylor, Victoria L. 2016. *The Brain's Body: Neuroscience and Corporeal Politics.* Durham, N.C.: Duke University Press.

Porter, Roy. 2006. *The Cambridge History of Medicine.* Cambridge: Cambridge University Press.

Rabinbach, Anson. 1992. *The Human Motor: Energy, Fatigue, and the Origins of Modernity.* Berkeley: University of California Press.

Read, Herbert. 1956. *The Art of Sculpture.* London: Faber.

Rey, Roselyne. 1995. *History of Pain.* Translated by L. E. Wallace, J. A. Cadden, and S. W. Cadden. Cambridge, Mass.: Harvard University Press.

Riegl, Aloïs. (1901) 1995. "Late Roman Art Industry." Translated by R. Winkes. In *Art History and Its Methods: A Critical Anthology,* edited by E. Fernie, 116–26. London: Phaidon.

Riegl, Aloïs. (1902) 1988. "Late Roman or Oriental?" In *German Essays on Art History: Winckelmann, Burckhardt, Panofsky and Others,* edited by G. Schiff, 173–90. New York: Continuum.

Rockefeller Foundation. 1930. *The Rockefeller Annual Report.* New York: Rockefeller Foundation.

Rosenberg, Martin, and Steve McMahon. 2004. "Extract from an Annotated Physiological Society Interview with Professor Patrick Wall (1925–2001)." In *Innovation in Pain Management.* Edited by L. A. Reynolds and E. M. Tansey, 73–82. Wellcome Witnesses to Twentieth Century Medicine, vol. 21. London: Wellcome Trust Centre for the History of Medicine at UCL.

Sachs, Carl. 1874. "Physiologische und anatomische Untersuchungen über die sensiblen Nerven der Muskeln." In *Archiv für Anatomie, Physiologie und Wissenschaftliche Medicin,* edited by C. B. Reichert and E. Du Bois-Reymond, 175–95. Leipzig, Ger.: Verlag von Veit et Comp.

Sacks, Oliver W. 1986. *The Man Who Mistook His Wife for a Hat and Other Clinical Tales.* London: Picador.

Salisbury, Laura, and Andrew Shail. 2010. "Introduction." In *Neurology and Modernity: A Cultural History of Nervous Systems, 1800–1950,* edited by L. Salisbury and A. Shail, 1–40. Cambridge: Palgrave.

Sample, Ian. 2014. "Shocking but True: Students Prefer Jolt of Pain to Being Made to Sit and Think." *Guardian,* July 3, 2014. https://www.theguardian.com/science /2014/jul/03/electric-shock-preferable-to-thinking-says-study.

Satz, Aura. 2010. "'The Conviction of Its Existence': Silas Weir Mitchell, Phantom Limbs and Phantom Bodies in Neurology and Spiritualism." In *Neurology and Modernity,* edited by L. Salisbury and A. Shail, 113–29. Cambridge: Palgrave.

Scarry, Elaine. 1985. *The Body in Pain: The Making and Unmaking of the World.* Oxford: Oxford University Press.

Schäfer, Karl. 1889. *Die Erklärung der Bewegungsempfindungen durch den Muskelsinn.* Jena, Ger.: Druck von B. Engan.

Schilder, Paul. (1935) 2013. *The Image and Appearance of the Human Body: Studies in the Constructive Energy of the Psyche.* London: Routledge.

Schiller, Francis. 1984. "Coenesthesis." *Bulletin of the History of Medicine* 58 (4): 496–515.

Schilling, Richard S. F. 1944. "Industrial Health Research: The Work of the Industrial Health Research Board, 1918–44." *British Journal of Industrial Medicine* 1 (3): 145–52.

Schivelbusch, Wolfgang. 1986. *The Railway Journey: Trains and Travel in the Nineteenth Century.* Oxford: Blackwell.

Schmidgen, Henning. 2014. *The Helmholtz Curves: Tracing Lost Time.* New York: Fordham University Press.

Schott, Geoffrey D. 1993. "Penfield's Homunculus: A Note on Cerebral Cartography." *Journal of Neurology, Neurosurgery, and Psychiatry* 56:329–33.

Scientific American. 1878. "A Horse's Motion Scientifically Determined." 39 (16): 241.

Scientific American. 1884. "The Motor Power of the Human Body." 51 (19): 290.

Scientific American. 1896. "Chronophotography." 75 (19): 344.

Scott, Geoffrey. 1914. *The Architecture of Humanism: A Study in the History of Taste.* New York: W. W. Norton.

Sheets-Johnstone, Maxine. 1992. "The Materialization of the Body: A History of Western Medicine, a History in Process." In *Giving the Body Its Due,* edited by M. Sheets-Johnstone, 132–58. Albany: State University of New York Press.

Sherrington, Charles S. 1900a. Letter to Acland, December 9, 1900. Item MS7623/2, Wellcome Library Archives.

Sherrington, Charles S. 1900b. "The Muscular Sense." In *Textbook of Physiology,* vol. 2, edited by E. A. Schäfer, 1002–25. Edinburgh: Young J. Pentland.

Sherrington, Charles S. 1903. "Qualitative Differences of Spinal Reflex Corresponding with Qualitative Difference of Cutaneous Stimulus." *Journal of Physiology* 30:39–46.

Sherrington, Charles S. 1906. *The Integrative Action of the Nervous System.* New Haven, Conn.: Yale University Press.

Sherrington, Charles S. 1907. "On the Proprio-Ceptive System, Especially in Its Reflex Aspect." *Brain* 29 (4): 467–82.

Sherrington, Charles S. 1933. *The Brain and Its Mechanism.* Cambridge: Cambridge University Press.

Sherrington, Charles S. 1940. *Man on His Nature: The Gifford Lectures, Edinburgh, 1937-8.* Cambridge: Cambridge University Press.

Shusterman, Richard. 2011. "Somatic Style." *Journal of Aesthetics and Art Criticism* 69 (2): 147–59.

Sinclair, Hugh M. 1984. "Sherrington and Industrial Fatigue." *Notes and Records of the Royal Society of London* 39 (1): 91–104.

Solnit, Rebecca. 2004. *River of Shadows: Edweard Muybridge and the Technological Wild West.* New York: Penguin.

Solomon, Joshua A. 2011. "Commemorating *Elemente der Psychophysik.*" In *Fechner's Legacy in Psychology: 150 Years of Elementary Psychophysics,* edited by J. A. Solomon, 1–7. Leiden, Neth.: Brill.

Spencer, Herbert. 1872. *The Principles of Sociology.* Vol. 2. 2nd ed. London: Williams and Norgate.

Stewart, Susan. 1999. "Prologue: From the Museum of Touch." In *Material Memories,* edited by Marius Kwint, Christopher Breward, and Jeremy Aynsley, 17–36. Oxford: Berg.

Stillman, Jacob D. B. 1882. *The Horse in Motion: As Shown by Instantaneous Photography.* Boston: James R. Osgood.

Sully, James. 1876. "Physiological Psychology in Germany." *Mind* 1 (1): 20–43.

Swanson, Larry W. 2014. *Neuroanatomical Terminology: A Lexicon of Classical Origins and Historical Foundations.* Oxford: Oxford University Press.

Swazey, Judith P. 1969. *Reflexes and Motor Integration: Sherrington's Concept of Integrative Action.* Cambridge, Mass.: Harvard University Press.

Taussig, Michael. 2007. "Tactility and Distraction." In *Beyond the Body Proper: Reading the Anthropology of Material Life,* edited by M. Lock and J. Farquhar, 259–65. Durham, N.C.: Duke University Press.

Taylor, Frederick W. 1912. *The Principles of Scientific Management.* New York: Harper and Brothers.

Thomas, Inigo. 2016. "The Chase." *London Review of Books* 38 (20): 15–18.

Thompson, Graves S. 1956. "We See by the Papers." *Classical Journal* 52 (2): 79–80.

Times. 1844. "Royal Academy of Painting," May 8, 1844. *The Times Digital Archive.* Accessed October 9, 2019. http://tinyurl.gale.com/tinyurl/Bsa9wl.

Titchener, Edward B. 1908. "The Tridimensional Theory of Feeling." *American Journal of Psychology* 19 (2): 213–31.

Torres-Oviedo, Gelsy, Erin Vasudevan, Laura Malone, and Amy J. Bastian. 2011. "Locomotor Adaptation." *Progress in Brain Research* 191:65–74.

Tousignant, Noémi. 2014. "A Quantity of Suffering: Measuring Pain as Emotion in the Mid-twentieth Century USA." In *Pain and Emotion in Modern History,* edited by R. Boddice, 111–29. New York: Palgrave Macmillan.

Treib, Marc. 2007. "Moving the Eye." In *Sites Unseen: Landscape and Vision,* edited by D. S. Harris and D. F. Ruggles, 61–88. Pittsburgh: University of Pittsburgh Press.

U.S. Bureau of Labor Statistics. 2012. "American Time Use Survey: 2012." www.bls.gov /tus/home.htm#data.

Verdin, Charles. 1890. *Catalogue des instruments de précision pour la physiologie et la médecine.* Paris: Imp. Mersch.

Vinge, Louise. 1975. *The Five Senses: Studies in a Literary Tradition.* Lund, Swe.: Royal Society for the Humanities.

Vitruvius. 1914. *The Ten Books on Architecture.* Translated by M. H. Morgan. Cambridge, Mass.: Harvard University Press.

Volkelt, Johannes. 1876. *Der Symbolbegriff in der neuesten Ästhetik.* Jena, Ger.: Hermann Dufft.

Von Uexküll, Jakob. (1932) 1992. "A Stroll through the Worlds of Animals and Men: A Picture Book of Invisible Worlds." *Semiotica* 89 (4): 319–91.

Wade, Nicholas J. 2003. "The Search for a Sixth Sense: The Cases for Vestibular, Muscle, and Temperature Senses." *Journal of the History of the Neurosciences* 12 (2): 175–202.

Walker, Robin, and Melanie Doyle. 2003. "Multisensory Interactions in Saccade Generation." In *The Minds Eye: Cognitive and Applied Aspects of Eye Movement Research,* edited by J. Hyönä, R. Radach, and H. Deubel, 89–102. Amsterdam: Elsevier.

Walshe, Francis M. R. 1943. Letter from F. M. R. Walshe to Penfield, April 25, 1943. Correspondence Wilder Penfield with Sir Francis Walshe, C/D 20, Wilder Penfield Fonds, Osler Library of the History of Medicine, McGill University.

Walshe, Francis M. R. 1953. "Some Problems of Method in Neurology," *Canadian Medical Association Journal* 68 (1): 21–29.

Walshe, Francis M. R. 1958. "Some Reflections upon the Opening Phase of the Physiology of the Cerebral Cortex, 1850–1900." In *The Brain and Its Functions,* edited by F. M. L. Poynter, 223–34. Springfield, Ill.: Charles C. Thomas.

Walshe, Francis M. R. 1961. "Contributions of John Hughlings Jackson to Neurology: A Brief Introduction to His Teachings." *Archives of Neurology* 5 (2): 119–31.

Ward, James. 1919. *Psychological Principles.* Cambridge: Cambridge University Press.

Webb, Jonathan. 2014. "Do People Choose Pain over Boredom?" BBC News, July 4, 2014. http://www.bbc.co.uk/news/science-environment-28130690.

Weber, Ernst H. (1846) 1905. *Tastsinn und Gemeingefühl.* Edited by E. Hering. Leipzig, Ger.: W. Engelmann.

Weber, Ernst H. 1978. *The Sense of Touch.* Translated by Helen E. Ross. New York: Academic Press.

Weber, Ernst H. 1996. *E. H. Weber on the Tactile Senses.* Translated by Helen E. Ross and David J. Murray. Hove, UK: Erlbaum, Taylor and Francis.

Wickens, Andrew P. 2014. *A History of the Brain: From Stone Age Surgery to Modern Neuroscience.* New York: Psychology Press.

Wiest, Gerald, and Robert W. Baloh. 2002. "The Pioneering Work of Josef Breuer on the Vestibular System." *Archives of Neurology* 59:1647–53.

Wilson, Timothy D., David A. Reinhard, Erin C. Westgate, Daniel T. Gilbert, Nicole Ellerbeck, Cheryl Hahn, Casey L. Brown, and Adi Shaked. 2014. "Just Think: The Challenges of the Disengaged Mind." *Science* 345 (6192): 75–77.

Winston, Brian. 2003. "Stopping Motion." *London Review of Books* 25 (15): 4.

Winston, Brian. 2005. *Messages: Free Expression, Media and the West from Gutenberg to Google.* New York: Routledge.

Wölfflin, Heinrich. (1886) 1993. "Prolegomena to a Psychology of Architecture." In *Empathy, Form, and Space: Problems in German Aesthetics, 1873–1893,* edited by H. F. Mallgrave, H. Francis, and E. Ikonomou, 149–87. Santa Monica, Calif.: Getty Research Institute for the History of Art and the Humanities.

Wölfflin, Heinrich. 1964. *Renaissance and Baroque.* Translated by K. Simon. London: Collins.

Wölfflin, Heinrich. 2003. "Linear and Painterly." In *The Visual Turn: Classical Film Theory and Art History,* edited by A. D. Vacche, 51–55. New Brunswick, N.J.: Rutgers University Press.

Woolf, Clifford J., and Quifu Ma. 2007. "Nociceptors—Noxious Stimulus Detectors." *Neuron* 55 (3): 353–64.

Wundt, Wilhelm M. (1863) 1896. *Lectures on Human and Animal Psychology.* Translated by J. E. Creighton and E. B. Titchener. New York: Macmillan.

Wundt, Wilhelm M. 1863. *Vorlesungen über die Menschen- und Thier-Seele* [Lectures on human and animal psychology]. 2 vols. Leipzig, Ger.: Voss.

Wundt, Wilhelm M. 1874. *Grundzüge der physiologischen Psychologie.* Leipzig, Ger.: Engelmann.

Wundt, Wilhelm M. (1896) 1902. *Outlines of Psychology.* Translated by C. H. Judd. Second Revised English Edition. New York: Gustav E. Stechert.

Wundt, Wilhelm M. 1897. *Outlines of Psychology.* Translated by C. H. Judd. Leipzig, Ger.: Wilhelm Engelmann.

INDEX

MARK PATERSON is associate professor of sociology at the University of Pittsburgh. He is author of *The Senses of Touch: Haptics, Affects, and Technologies*; *Seeing with the Hands: Blindness, Vision, and Touch after Descartes*; two editions of *Consumption and Everyday Life*; and coeditor of *Touching Space, Placing Touch*.